Building an Enterprise Chatbot

Work with Protected Enterprise Data Using Open Source Frameworks

Abhishek Singh
Karthik Ramasubramanian
Shrey Shivam

Apress®

Building an Enterprise Chatbot: Work with Protected Enterprise Data Using Open Source Frameworks

Abhishek Singh
New Delhi, Delhi, India

Karthik Ramasubramanian
New Delhi, Delhi, India

Shrey Shivam
Donegal, Donegal, Ireland

ISBN-13 (pbk): 978-1-4842-5033-4
https://doi.org/10.1007/978-1-4842-5034-1

ISBN-13 (electronic): 978-1-4842-5034-1

Managing Director, Apress Media LLC: Welmoed Spahr
Acquisitions Editor: Celestin Suresh John
Development Editor: Matthew Moodie
Coordinating Editor: Aditee Mirashi

Cover designed by eStudioCalamar

Cover image designed by Freepik (www.freepik.com)

Distributed to the book trade worldwide by Springer Science+Business Media New York, 233 Spring Street, 6th Floor, New York, NY 10013. Phone 1-800-SPRINGER, fax (201) 348-4505, e-mail orders-ny@springer-sbm.com, or visit www.springeronline.com. Apress Media, LLC is a California LLC and the sole member (owner) is Springer Science + Business Media Finance Inc (SSBM Finance Inc). SSBM Finance Inc is a **Delaware** corporation.

For information on translations, please e-mail rights@apress.com, or visit www.apress.com/rights-permissions.

Apress titles may be purchased in bulk for academic, corporate, or promotional use. eBook versions and licenses are also available for most titles. For more information, reference our Print and eBook Bulk Sales web page at www.apress.com/bulk-sales.

Any source code or other supplementary material referenced by the author in this book is available to readers on GitHub via the book's product page, located at www.apress.com/978-1-4842-5033-4. For more detailed information, please visit www.apress.com/source-code.

Printed on acid-free paper

Abhishek and Karthik dedicate this book to their parents for their unwavering support and love.

Shrey dedicates this book in memory of his grandparents, the late Mr. Ravindra Narayan Singh and late Dr. Ganga Prasad Singh, for being the source of his inspiration and pride.

Table of Contents

About the Authors

Abhishek Singh is on a mission to profess the de facto language of this millennium, the numbers. He is on a journey to bring machines closer to humans, for a better and more beautiful world by generating opportunities with artificial intelligence and machine learning. He leads a team of data science professionals solving pressing problems in food security, cyber security, natural disasters, healthcare, and many more areas, all with the help of data and technology. Abhishek is in the process of bringing smart IoT devices to smaller cities in India so that people can leverage technology for the betterment of life.

He has worked with colleagues from many parts of the United States, Europe, and Asia, and strives to work with more people from various backgrounds. In seven years at big corporations, he has stress-tested the assets of U.S. banks at Deloitte, solved insurance pricing models at Prudential, made telecom experiences easier for customers at Celcom, and created core SaaS data products at Probyto. He is now creating data science opportunities with his team of young minds.

He actively participates in analytics-related thought leadership, authoring, public speaking, meetups, and training in data science. He is a staunch supporter of responsible use of AI to remove biases and fair use of AI for a better society.

Abhishek completed his MBA from IIM Bangalore, a B.Tech. in Mathematics and Computing from IITGuwahati, and has a PG Diploma in Cyber Law from NALSAR University, Hyderabad.

Karthik Ramasubramanian has over seven years of practice in data science and business analytics in retail, FMCG, e-commerce, and information technology for a multi-national and two unicorn startups. He is a researcher and problem solver with a diverse set of experiences in the data science lifecycle, starting from a data problem discovery to creating a data science prototype/product.

On the descriptive side of data science, he designed, developed, and spearheaded many A/B experiment frameworks for improving product features, conceptualized funnel analysis for understanding user interactions and identifying the friction points within a product, and designed statistically robust metrics and visual dashboards. On the predictive side, he developed intelligent chatbots which understand human-like interactions, customer segmentation models, recommendation systems, identified medical specialization from a patient query for telemedicine, and other projects.

He actively participates in analytics-related thought leadership, authoring blogs and books, public speaking, meet-ups, and training and mentoring for data science.

Karthik completed his M.Sc. in Theoretical Computer Science at PSG College of Technology, India, where he pioneered the application of machine learning, data mining, and fuzzy logic in his research work on the computer and network security.

Shrey Shivam has extensive experience in leading the design, development, and delivery of solutions in the field of data engineering, stream analytics, machine learning, graph databases, and natural language processing. In his seven years of experience, he has worked with various conglomerates, startups, and big corporations, and has gained relevant exposure to digital media, e-commerce, investment banking, insurance, and a suite of transaction-led marketplaces across the music, food, lifestyle, news, legal, and travel domains.

He is a keen learner and is actively engaged in designing the next generation of systems powered by artificial intelligence-based analytical and predictive models. He has taken up various roles in product management, data analytics, digital growth, system architecture, and full stack engineering. In this era of rapid acceptance and adoption of new and emerging technologies, he believes in strong technical fundamentals and advocates continuous improvement through self-learning.

Shrey is currently leading a team of machine learning and big data engineers across the U.S., Europe, and India to build robust and scalable big data pipelines to implement various statistical and predictive models.

Shrey completed his BTech in Information Technology from Cochin University of Science and Technology, India.

About the Technical Reviewer

Santanu Pattanayak currently works at GE, Digital as a Staff Data Scientist and is the author of *Pro Deep Learning with TensorFlow - A Mathematical Approach to Advanced Artificial Intelligence in Python*. He has 12 years of overall work experience with 8 years of experience in the data analytics/data science field. He also has a background in development and database technologies. Prior to joining GE, Santanu worked at companies such as RBS, Capgemini, and IBM. He graduated with a degree in Electrical Engineering from Jadavpur University, Kolkata and is an avid math enthusiast. Santanu is currently pursuing a master's degree in Data Science from Indian Institute of Technology (IIT), Hyderabad. He also devotes time to data science hackathons and Kaggle competitions where he ranks within the top 500 across the globe. Santanu was born and brought up in West Bengal, India and currently resides in Bangalore, India with his wife.

Acknowledgments

We are grateful to our teachers at various universities and their continued support in our professional lives.

Abhishek Singh thanks his colleagues at Probyto who inspire him to write impactful content for better use of AI for public use; the idea of this book evolved through discussions with his colleagues and his work in the EU market. A special mention goes to his parents, Mr. Charan Singh and Mrs. Jayawati, for their intriguing insights on how to think about general human use of AI. Their support and demand for the simplistic design of solutions using AI-generated data inspires his day-to-day design of data products.

Karthik is immensely grateful to his parents, Mr. S Ramasubramanian and Mrs. Kanchana, for their unwavering support throughout his life and during the development of this book. This book was made possible by hundreds of researchers who shared their life's work as open-source offerings. He thanks all such researchers who make this world better and passionately share their work with everyone. Lastly, a large part of his work and success comes from his mentors and colleagues from work, where he continuously learns and improves.

Shrey is hugely grateful to his parents, Mr. Vijay Pratap Singh and Mrs. Bharti Singh, for their love, care, and sacrifice in helping him fulfill his dreams. He expresses gratitude to his uncle, Mr. Tarun Singh, for being a pillar of strength. Shrey also thanks his past and current colleagues, including Dipesh Singh and Jaspinder Singh Virdee, for their continuous encouragement and support in taking up challenging and innovative ideas to execution.

ACKNOWLEDGMENTS

Finally, this book would not have been possible without the support of the Apress team: Aditee, Celestin, Matthew, and the production support staff. We also acknowledge and thank our Technical Reviewer (TR) for their critical reviews that helped to make the book even better.

Introduction

There are numerous frameworks and propriety off-the-shelf chatbot offerings, but most do not clearly map out the much-needed control of data by an organization. Often the propriety offerings take an organization's private data for training on the cloud and provide the outcome as a model. In this book, we will focus on data privacy and control over the development process. The chatbot that you will learn how to develop could be completely built in-house using open-source JAVA frameworks and NLP libraries in Python.

The beginning of the book helps you understand the processes in the banking industry and delves into identifying the sources of data to mine for the intent from customer queries. The second part focuses on natural language processing, understanding, and generation, which are demonstrated using Python. These three concepts are the core components of a chatbot. In the final part, you take up the development of a chatbot called IRIS using open-source frameworks written in JAVA.

The following are the key topics this book offers:

- Identify the business processes where chatbots could be used in an industry and suitably guide the design in a solution architecture

- Focus on building a chatbot for one industry and one use-case, rather than building a ubiquitous and generic chatbot

- Natural language understanding, processing, and generation

- Learn how to deploy a complete in-house-built chatbot using an open source technology stack like RASA and Botpress (such chatbots avoid sharing any PIIs with any third-party tools)

- Develop a chatbot called IRIS from scratch by customizing an existing open-source chatbot framework

- Use APIs for chatbot integration with internal data sources

- Deployment and continuous improvement framework through representational learning

We hope you enjoy the journey.

CHAPTER 1

Processes in the Banking and Insurance Industries

According to Darwin's *On the Origin of Species*, it is not the most intellectual of the species that survives; it is not the strongest that survives; the species that survives is the one that is best able to adapt and adjust to the changing environment in which it finds itself. The same analogy can apply to enterprises and their survival opportunities in the 21st century. In this digital era, it is of utmost importance for enterprises to adapt to the latest trends and technology advancements. With this book, we intend to prepare you with an emerging skill of building chatbots in the financial services domain, with a specific use case of an insurance agent (replicable to a bank assistant as well).

Banking and Insurance Industries

Banks and insurers have been in existence for a long time and facilitate economic activities for us. Banking and insurance play essential roles in the economic growth of a country and society. Both institutions provide the essential services of commercial transactions and covering risks.

© Abhishek Singh, Karthik Ramasubramanian, Shrey Shivam 2019
A. Singh et al., *Building an Enterprise Chatbot*,
https://doi.org/10.1007/978-1-4842-5034-1_1

Insurance services evolved from the practice of risk management for uncertain events. The risk is defined as the uncertainty of outcome in the normal process. The risks are quantified in monetary terms with consequences that are unfavorable to the process. The insurance function tends to manage the risk by providing a security net against a payment. In financial terms, insurance transfers the risk of the unfavorable event to the insurer against a payment of a premium.

As seen in Figure 1-1 , the primary function of an insurance company is to manage the fund created by the premiums paid by the insured. The critical function of an insurance company is to measure the risk of loss arising from the pool to decide on premiums and, in case of an accident/ adverse outcome, pay the policyholder the loss amount. As the count of adverse outcomes decreases with an increasing population, smaller premiums can be levied while a higher payout can be made to the insured who faces an adverse outcome.

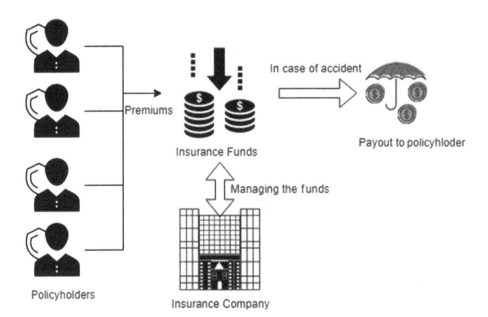

Figure 1-1. Theoretical framework of insurance

The insurance industry is comprises of evolved financial services and products which are centuries old. With the advent of technology, the insurance industry has seen a surge in number of big insurance companies and deeper penetration with new products. Having a concentration brings better premiums for the insured and allows companies to cover a broader set of risks. As seen in Figure 1-2, typically the insurance products can be divided into two categories: life insurance and general insurance.

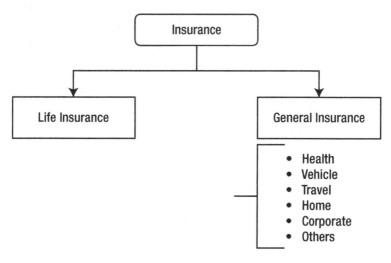

Figure 1-2. *Insurance product categorization*

Life insurance provides coverage to the risk of mortality. The insured beneficiary receives a face amount in case of death during the coverage period. In this manner, this insurance product safeguards against financial losses arising from the death of a critical member of the family. The risk covered is called mortality risk. The actuarial science is the study of mortality behavior and used in study of fair pricing of premiums for the given subject for life insurance.

Insurance is not just limited to insuring against the risk of death. The insurance concept has been extended to other forms of risk as well. The other bucket of coverage is known as general insurance; this includes health insurance covering the financial risk of ill health, vehicle insurance for accidents, travel insurance for flight delays, and so on. The established insurance companies offer multiple products to customers and institutions as per their needs. Some of these products have standard features, where some insurance companies can create custom deals as per client needs, for instance covering the risk of severe weather during a significant event. The key differentiator between a good and bad insurer is how diligently and accurately it can measure the risk involved in the underlying events.

Banking services, on the other hand, do not cover uncertain risk, but they work in economic activities of a financial nature. Banking has also evolved to be of many types, serving different purposes for different commercial entities. However, the basic premise of banking remains as a broker between lenders and borrowers. The spread between lending and borrowing rates is also called a spread, and the bank manages to create economic activities in the system.

Figure 1-3 depicts a fundamental framework of a bank or a banking company.

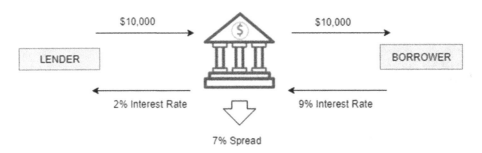

Figure 1-3. *Theoretical framework of banking*

Lenders have access to capital, such as institutional members having excess cash or a small retail customer who has some savings. The borrower is short of capital, but they have some economic activities which can bring returns on their investment. The bank comes into the play to solve this gap in the financial system and creates an opportunity for the lender to earn interest on deposits and lets the borrower get the required capital for an interest rate. In the example in Figure 1-3, a lender deposits $10,000 into the bank and receives 2% interest (i.e., $200), while the bank lends $10,000 to the borrower at 9% interest (i.e., gaining $900 in the transaction). The spread of 7% (i.e., 900-200=700) is the income for bank, which it can use to run operations and create new products.

Like insurance, banks have also evolved to provide various services for different types of customers and entities. Figure 1-4 is basic classification of types of banks. It is a not an exhaustive list of the types of banks and banking services. However, they are the primary type of banks.

Figure 1-4. *Common types of banks*

Within the scope of this book, we will point out the typical process for an end customer for accessing financial service. The customer for a retail bank and life insurance company are the same, except for a few cases. This makes it easier to illustrate how insurance company touchpoints with customers are similar to those for banking customers. Once we set up the generic nature of these touchpoints, we will move ahead with the chatbot build process.

In retail banking, an end customer, usually an entity or individual, deposits savings in a bank, and other entities or individuals borrow that money for other purposes. Apart from that, banks also facilitate online transactions, paying billers, transferring money to other entities, timed deposits, and many other services for retail customers.

A Customer-Centric Approach in Financial Services

Customer behavior and interactions have evolved to a personalized approach over the last two decades. Competition and greater reliance on technology for delivery of services are keys to this change in customer behavior. The importance of a customer-centric approach in products and services is far greater than ever before. The customer-centric approach involves many direct and indirect interventions through multiple channels (see Figure 1-5).

Figure 1-5. *Customer centricity in financial services*

The core element of customer centricity is a focused customer leadership. If the leadership aligns the strategy to become a customer-centric organization, the whole outlook and communications become customer-centric. Amazon has proved this and is now seen as the benchmark for a customer-centric approach. Understanding the customer and designing experiments to validate the hypothesis form the next steps in a customer-centric approach. Once we set up a successful connection with customers, we need to empower the front line, track essential metrics, and keep the feedback cycle. These are some indicative steps to achieve a customer-centric approach.

In financial services, primarily referring to retail products/services, the interaction points are many, and all touchpoints are critical to being customer focused. Banks and insurance companies deal with many individuals customers daily via multiple channels.

Digital interventions are redefining the ways customer engagement happens. There are some critical trends among customers accessing banking and insurance services.

- **More natural interactions**: The user experience is of the utmost importance. The customers are looking for easier access to products, an appealing experience, and easy action in a few clicks.

- **More touchpoints and flexibility**: The customer does not want fixed 9 am to 5 pm branch visits or no access on weekends. Customers want to be able to access and buy products anytime and via multiple channels. It may be a mobile app or a web app or phone banking, but they want more flexibility in how they interact.

- **Responsive service**: Customers expect that the bank/ insurer knows about them and is responsive to their needs. They want individual attention and appreciate responsive customer service.

- **Clear product information**: With so many players and products, customers want concise and relevant information to be delivered to them. Additional details they can seek with follow-ups. The customer does not want a pile of information or to get confused.

- **Great value from the products**: The product features are numerous and many times the customer is unaware of how to make the best use of them. Customers expect the bank/insurer to keep reminding them to draw the best value out of a product and if possible, offer new products that might be useful.

The growing digital presence of financial institutions also requires multiple changes in the technology landscape. Traditional database systems and applications are now becoming obsolete. Powerful endpoint computing (i.e., smartphones), excellent internet connectivity (i.e., 4G/5G), and cloud platforms are the magic trinity for a digital revolution in the financial sector.

In this book, we will explore the evolution and working of chatbots in many endpoint interactions with customers. While conversational agents have existed for a long time (remember the IVRS systems?), new technology developments have made them driven by natural language, offering customer-centric delivery of information. Chatbots are designed to carry out specific and structured interactions; the complex service interactions are still better served by an experienced customer service representative. In coming chapters, we will cover different aspects of building a chatbot for an insurance agent.

Benefits from Chatbots for a Business

According to the Grand View Research 2018 report,[1] the global chatbot market is expected to reach $1.25 billion by 2025, with a CAGR of 24.3% (average annual growth rate). The chatbot market will grow significantly across the financial services sectors, as they are among the largest customer-facing businesses (in our context, the insurance business). The immediate value creation for institutions happens by significantly reducing the operating cost and bringing customer satisfaction.

Technically speaking, chatbots are a combination of technology, artificial intelligence (AI), and business process designs.

- The technology provides the carrier for exchanging messages between chatbots and customers, and chatbots and internal systems, delivering information in real time over mobile phone or the Web.

- The AI builds the core brain of the chatbot, which understands the natural language decoded from machine instructions. They also make decisions during conversations.

- The most critical piece is the business process design, which identifies the standard process to access information, what information can be shared with whom, and convenient ways to buy/sell/inquire about current products.

While chatbots offer immense monetary value for the company regarding reducing the cost of customer service and as a new channel for revenue by sales of products and services, they also add immense value to the customer's experience.

---

[1] www.grandviewresearch.com/press-release/global-chatbot-market

- **24x7 availability**: Chatbots are available 24x7 through phones or the Web. This gives the customer options of when to interact with the services.

- **Zero human touch experience**: Chatbots allow customers to have a zero human touch experience for their basic requirements. This way of getting the necessary information without going through the manual route is entirely new.

- **Simplicity**: Chatbots simplify the process for customers by decoding the process into clear steps. The information delivered by chatbots is also very concise and to the point, as per the customer query.

Daily, the ever-changing chatbot market is coming up with disruptive ideas and delivering value across the spectrum.

Chatbots in the Insurance Industry

Gone are the days of waiting for the next available operator or taking the effort to get the information. The customer-centric approach is one of the critical differentiators for a company today. Chatbots are helping augment customer engagement and brand presence, and they are proving to be very useful in most industries including the insurance industry. The emergence of mobile and social media has not only provided new channels of communication between people but has also made people feel closer to businesses. Companies are investing heavily in creating and maintaining a robust digital presence and implementing new solutions so they can have a better customer reach.

Traditionally, the insurance industry has been slow to change. Due to the complexity of insurance, covering a diverse set of risks, the operating model has been cumbersome, with a lot of paperwork, background checks, and approvals. With the new era of digital business and increasing competition, the insurance industry is also addressing the needs of the always-on, always-connected digital world.

Roughly 70% of calls to a call center of an insurance company are queries that can be addressed without a human interface, such as customers requesting details on their claim status, policy renewal, or information on financial advisors. According to World Wide Call Centers,[2] in a shared call center, an inbound call rate ranges from $.35-$.45/minute at low-cost international agencies to $.75-$.90/minute in the U.S./Canada and from $8-$15 internationally to $22-$28 in the U.S./Canada. A typical large insurance company gets more than 10 million calls every year. Considering a call rate of $5 per call, even if a chatbot can address half of these queries, that's a potential savings of $25 million per year.

Where live agents can handle only two to three conversations at a time, a chatbot can operate without any such limit and reduce the human resources required to handle such queries. It can also automate repetitive work. These calls to call centers have an average wait time of 3 minutes until they are assigned to an agent, and customers who browse the website typically spend around 5 to 10 minutes to find the information they require. Virtual agents such as chatbots provide this information in real time, which is significant use of technology to make the interaction faster and more efficient. The services can be accessed 24x7 through multiple digital interaction platforms such as mobile apps, Facebook Messenger, Twitter, SMS, Skype, Alexa, and web UI chat, providing omnichannel support to customers.

Figure 1-6 shows the ways in which chatbots are transforming the insurance industry.

---

[2]www.worldwidecallcenters.com/call-center-pricing/

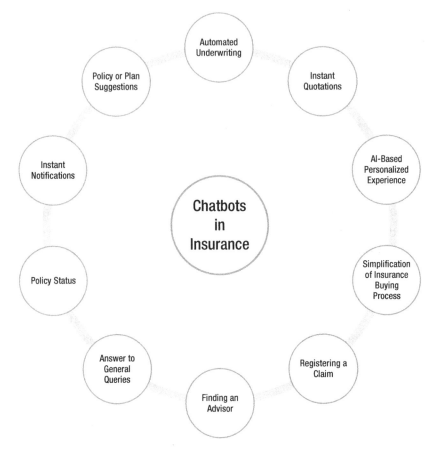

Figure 1-6. *Chatbots in the insurance industry*

Some of the most common applications of a chatbot in the life insurance industry are in the following sectors.

Automated Underwriting

With a wealth of information available about individuals online, machine learning techniques are being used to assess the risk index of an individual accurately. Companies are using virtual digital agents (chatbots) to provide a simplified way of buying life insurance and getting an instant decision.

Instant Quotations

Customers can get insurance eligibility and quotation details on the platform of their choice instantly.

AI-Based Personalized Experience

Since chatbots are designed to simulate human interaction, they can leverage AI to understand context and user needs in order to provide a better customer satisfaction experience.

Simplification of the Insurance Buying Process

The general public has an aversion to insurance-related paperwork due to long forms that are difficult to understand. Chatbots can ask the customer simple questions in a conversational language and use the answers to auto-populate some of the fields on the online form, speeding up the application process.

Registering a Claim

Since chatbots are virtual agents, they are available 24×7. Hence, they can help customers with the claim process regardless of the time of the incident. Time-to-settle-claim is an important metric that plays a critical role in improving the efficiency of an insurance business, and chatbots are playing an essential role in reducing the overall time to settle.

Finding an Advisor

Some companies use insurance agents or advisors. The natural language-based interaction model of the chatbots makes them convenient for customers to quickly inquire about insurance agents or financial advisors based on location or insurance type.

Answering General Queries

As many as 30% of the queries to the call center are general queries asking about policy cash value, policy premium due date, interest rates, FAQs, company and product information, account-related issues such as password reset, updating the beneficiary, and details around the application process. These queries can be addressed by a chatbot in real time in a customer-centric way.

Policy Status

The customer can check policy status and statuses of claims or other complaints or requests by interacting with a chatbot anywhere anytime.

Instant Notifications

Chatbots can remind customers about the policy premium due date, next billing cycle, and so forth.

New Policy or Plan Suggestions

Chatbots not only perform the role of service agents but also provide new marketing opportunities. User interaction and social media behavior on digital platforms such Facebook can be tailored to suggest content, products, or services as per their needs.

Conversational Chatbot Landscape

In this internet era, every time a person requires a service or information, the person must find an appropriate website. In the mobile era, native apps took center stage with the same purpose as a website. Every business nowadays has one website and one mobile app at the bare minimum. Now, in the AI era, a customer is flooded with information on the website and the mobile app, and there aren't as many employees to help the many customers who are seeking the service or information. Moreover, even if a company finds many employees, the cost is very high. Conversational bots or chatbots are playing a pivotal role in the AI era by addressing the critical problem of information deluge at an affordable cost.

The organization is going through a digital transformational journey where chatbots are being discussed in roadmaps. The primary objective is to improve customer experience through simplified touchpoints and faster service time. This objective often results in higher conversion for newer products and services and reduced cost of operations.

A massive growth of bot frameworks (technology) and advancements in the natural language understanding (AI) has led to the adaption of chatbots in many industries. Companies are building chatbots across the lifecycle of their customers, namely

- Acquisition

- Engagement

- Servicing

- Feedback

Acquisition and engagement help companies build a strong top line for the business while servicing helps in reducing cost and feedback increases customer retention.

Figure 1-7 shows an industry-wide adaption of chatbots and various use-cases.

Insurance

- Claim and Coverage Queries
- Underwriting
- Automated Advisory

Financial Services

- Account Management
- Product Guidance
- Predictive Offers
- Market Updates
- Online Subscription Managements

Healthcare

- Discovery and Scheduling
- Care Management
- Drug Information
- Diagnostic Test Appointments
- Assist Non-Emergency Patient Queries

Travel

- Automatic Flight Reminders and Updates
- Virtual AI Travel Agent
- Book Hotels, Cabs, and Restaurants
- Suggest Sight-Seeing and Restaurants

Retail

- Product Search in Super Marts
- Recipe Search
- Locate Nearby Stores
- Order Food

Telecommunications

- Billing and Accounts Services
- Offers and Plan Changes
- Customer Support and Self-Service
- Training and Operations Productivity

Figure 1-7. *Conversational chatbot landscape*

In terms of benefits, insurance companies are now able to process their claims 24% faster,[3] and telcos are achieving as much as 90%[4] reduction in their customer service calls, in which 75% is reduced by self-service guides and automatic tips and an additional 15% of calls with the help of an artificial intelligence agent. This leaves only 10% to the costly phone call operator. By answering up to 80% of routine questions in a service center, the customer service cost is reduced by 30%[5] and companies like Autodesk saw 99%[6] improvement in response time for their level 1 queries. The benefits are continually increasing with more adaption and improvements.

In the upcoming chapters, we will delve into the details of NLU and various technologies to build a fully functional enterprise-grade chatbot. In the next chapter, we will discuss how to identify the customer interactions points, the data collection strategy, the importance of being compliant with privacy laws and understanding the data flow for each interaction with the client.

Summary

This chapter focused on processes in the banking and insurance industry where chatbots are bringing a new wave of innovation. Tasks that were earlier thought to be possible only by humans are now getting automated.

---

[3]www.avaamo.com/

[4]www.mckinsey.com/industries/telecommunications/our-insights/a-future-for-mobile-operators-the-keys-to-successful-reinvention

[5]https://chatbotsmagazine.com/how-with-the-help-of-chatbots-customer-service-costs-could-be-reduced-up-to-30-b9266a369945

[6]www.ibm.com/blogs/watson/2017/10/how-chatbots-reduce-customer-service-costs-by-30-percent/

Such innovation brings the cost down and helps in achieving scale. We also discussed chatbots in various other industries including healthcare and travel. Various industry reports were highlighted to prove the benefits of using AI-driven chatbots in an industry. In the coming chapters, we will build a conversational chatbot from scratch, keeping the focus on banking and insurance.

## CHAPTER 2

Identifying the Sources of Data

Chatbots are one more channel of providing conversational flows to customers. In the previous chapter, we discussed how the banking and insurance industries are structured and what kinds of interactions happen with the customers in those industries. There are many types of touchpoints a bank or insurer provides to customers in the day-to-day operations, starting from selling a new policy to settling escalations of claims. All these touchpoints are sources of data for building an AI Assistant, i.e., chatbot. In this chapter, we will start by introducing chatbot types and sources of data for training chatbots and then we will introduce the General Data Protection Regulation (GDPR) in context of the chatbot for personal data.

Chatbot Conversations

The chatbot tries to mimic the conversation of a real human. In the context of interacting with a business, the conversations can be of a broad, generic subject or particular to the product or service. Based on the scope of conversation, we can divide the conversation into two types: general conversations and specific conversations. The type of conversation decides

© Abhishek Singh, Karthik Ramasubramanian, Shrey Shivam 2019
A. Singh et al., *Building an Enterprise Chatbot*,
https://doi.org/10.1007/978-1-4842-5034-1_2

the scope of questions and knowledge the chatbot or human assistant needs in order to interact with a customer.

General Conversations

A general conversation is a typical conversation that happens when a customer and assistant are not confined to a specific topic or concern. The conversations can start from any point and can transverse to any direction based on the knowledge level of an assistant.

An example of such a conversation is

- A user walks into the bank and wants to talk to the manager. Before the start of the conversation, we don't know who this person is and why he is visiting the manager. The conversation can be about a sponsorship event, a loan account or utility payment, or something else.

To deal with such conversations, the chatbot needs to be built with many types of contexts and appropriate replies. The replies for such general conversations are also not heuristic in nature; they involve human natural intelligence and information/experience that is not available for the chatbot to build a conversation.

Many advances are happening in AI; we try to mimic complete human behavior by training with massive datasets and scenarios. However, we are still far away in terms of having complete general-conversation-based chatbots for industrial use cases.

Specific Conversations

Specific conversations are limited to some predesigned outcomes. These types of conversations have higher clarity of the scope of the talk and clear instructions to fall back on or cross-references to other sources.

Any deviations from the set conversations are generally directed to predefined outcomes. All other cases are redirected to appropriate channels, or the conversations ends.

An example of such a conversation is

- A customer walks into a store and goes to the refund desk. In this case, the refund desk has some specific conditions to process a refund and maybe some other particular functions. The customer cannot expect any other query than a refund to be offered at the refund desk. If he asks a question regarding discounts, he is directed by the refund desk to another counter.

Specific conversations are more predictable and can be handled with higher accuracy. The chatbots designed for specific tasks can communicate with information. The conversations are outcome-oriented and end once the outcome is achieved.

Training Chatbots for Conversations

Chatbots need to be taught how to have a conversation. The training of chatbots involves exposing chatbots to both rules and natural conversations. For general conversation chatbots, the amount of training data required is enormous, and so far we have not succeeded in creating an accurate general conversation chatbot. Alexa, Siri, and Google Home are few examples in this direction.

Creating chatbots also requires a set of rules documented or tacit to proceed with conversations. For example, if a chatbot asks for a customer's name, it must expect a first name and a last name. If the last name is not captured, it must go back and confirm the name. This is important to make sure the conversation is specific to the correct customer.

To train a chatbot for conversations, we need to have a corpus of training data. The training data can be accessed from multiple sources based on the use case. In the following sections, we discuss some datasets for use in training chatbots.

Self-Generated Data

Chatbot developers need to start with some data to make the chatbot come alive. Usually, that data is generated by developers for some necessary flow themselves. This way they get some high-level flow defined by themselves so that they can keep developing the chatbot with assumptions.

In many cases, developers create multiple inputs and self-annotate them for training basic flows; being generated by developers for testing the flows, they are not the complete set for training. These inputs help developers get the chatbot ready for a beta release and collect data from real users. Self-generated data is only a way to start development; it's not for general public use.

The data generated by developers is used to establish the data pipelines and system integration testing. Once the beta is deployed, the internal users can be exposed to the chatbot and more data is collected to keep training the natural language module.

Note   Do not confuse self-generated data with natural language generative (NLG) models. You will learn more about natural language generation from the small dataset in Chapter 5.

Customer Interactions

Customer interaction is the best source for training the chatbots. These conversations are the best to mimic for mainly two reasons:

- Typical queries can be captured, and chatbot training can be prioritized for specific conversations.

- The conversations capture real solutions provided in the past by experienced customer representatives.

Customer interactions happen through multiple channels, and these channels produce data for training chatbots as a new channel for customer interactions. Figure 2-1 shows the six main types of customer interaction channels for any modern business, applicable to our case of an insurance and bank as well.

Figure 2-1. *Customer interaction/service channels*

Phone

Phone calls are attended by experienced call center representatives and mostly accessed when the customer requires an immediate resolution to their queries. In modern days, this mode is recommended as the last step since it is costly for companies to maintain.

From the call center, we can get call transcripts, call recordings, core issues, and their resolutions. Core issues identified during calls and their resolution can help our chatbot learn to identify issues and provide solutions.

Emails

Email conversations are usually detailed and have a chronology of events explained and a clear statement of what the customer wants. These emails can be a good source to capture issues that need more than one-dimensional data to solve them.

Customer email records can be accessed in plain text format, with original emails and the response trail to developed conversations.

Chat

Many financial institutions use online web chat with customer service representatives to make sure they can serve multiple clients at the same time and reduce dropout of incoming queries at the call center.

This data set is very close to what a chatbot needs to mimic a conversation. Past chatlogs can be accessed as plain text files.

Social Media

Social media become popular when social media companies allowed business accounts to be created on their platforms. The interaction of social media tends to be generic and difficult to track with the actual customer of the general population.

Some platforms allow business accounts to download their data while some allow extracting data from API endpoints.

Customer Self-Service

Some necessary troubleshooting processes are created as self-service portals for customers. They may be as trivial as changing the PIN or offering FAQs for more information. Successful self-service cases are good for creating processes to train the chatbot to help people who ignore or can't use self-service.

This data is usually structured as a tree of conversations leading to the solution of specific problems.

Mobile

Mobile here is considered the interactions that happen via mobile apps and mobile browsing history by customers. The data captured from these mobile applications is captured as activity logs of customers.

Customer Service Experts

Customer service experts play a significant role in identifying typical customer queries and how they handle them in real situations. Their inputs are also helpful in creating default replies and designing fall-over options. The years of experience dealing with customers can be used to train as well as test the initial chatbot release.

Experts need to be part of the process of developing chatbots for quality assessment of the chatbots' experience and accuracy.

Open Source Data

Open source data is instrumental when you want to create general conversation chatbots and want to include some general flavor for specific talk. There are plenty of data sources available for training chatbots in natural language conversations.

A few of the open data sources are listed below; you can have more datasets as per your need.

- Yahoo Language Data, created from Yahoo Answers (`www.cs.cmu.edu/~ark/QA-data/`)

- WikiQA corpus, created from bing queries that redirect to wiki pages with a solution (`http://research.microsoft.com/apps/mobile/download.aspx?p=4495da01-db8c-4041-a7f6-7984a4f6a905`)

- Ubuntu Dialogue corpus, created from Ubuntu technical support (`www.kaggle.com/rtatman/ubuntu-dialogue-corpus`)

- Twitter data on Kaggle, created from customer support at Twitter (`www.kaggle.com/thoughtvector/customer-support-on-twitter`)

Crowdsourcing

The most critical training data comes from REAL people interacting with your chatbot in real time. This not only helps in building the corpus for training but also help developers see darker zones where the chatbots fail.

In best practice cases, all chatbots released at beta version are exposed to real conversations with selected customers and internal employees. The data is collected, and NLP models are retrained for each real instance. Another outcome of crowdsourcing is laying down the guidelines and scope of the chatbot.

Customer service experts also use the crowdsourcing inputs to build response languages and intensity for different conversations.

If you are building a chatbot in a regional language, you need to rely on crowdsourcing of training data. Some companies can provide you access to people who will interact with your chatbot to build the training corpus.

Personal Data in Chatbots

When we try to emulate human-like conversations with chatbots, we allow the humans to reveal information about themselves to the chatbot machine. This information then becomes risk for unauthorized access and may violate privacy laws and terms. This concern is of the utmost importance when you deal with customer queries that connect them to internal databases and required customer-specific information to process requests.

The customer can reveal the personal data both intentionally and unintentionally:

- Intentionally: To get an account balance, you need to provide an account number and PIN.

- Unintentionally: To know the claim process, you may end up revealing your policy number.

In both cases, the data is being captured by the chatbot, and the chatbot engine tries to process that data. Even if the chatbot can't process the data, it still creates a copy of a conversation that contains private and personal data of customers.

Another area where we expose personal data to our chatbots is at the time of training the chatbot. Internal data of customers might have personal, financial, and demographic information without the developer's full knowledge. For example, an email conversation regarding a claim settlement will contain a lot more details than just the customer-agnostic settlement process.

In deployment and training, personal data is captured and is vulnerable for law infringement and hacking, but this data is important for developing custom-centric chatbots. If we do not capture the data, we will not be able to design a chatbot that can take actions and provide information from internal databases.

We require more information than a normal conversation to be able to develop a chatbot that can access customer data and provide real-time information, securely and privately. The personal information helps in developing

- Authentication and access

- Compliance to company policies

- A customer information retrieval system

- A third-party API retrieval system

There are other related services and databases that require personal information to allow access to customer information in the private data zone.

As we just explained, we need personal data and other private data from customers to make the 24x7 AI assistant function with relevant data. This requires us to be very sure of both the customer agreements and local/international data regulations. Complying with regulations becomes of the utmost importance for banks and insurance companies to build specific conversation chatbots.

This is a challenge for companies because it limits the companies from using well developed, third-party chatbot services like Alexa, Dialogflow, and Watson. These services require the data to be sent to their server and stored for chatbot conversations. The limitations have created a vacuum to be filled by frameworks that can develop state-of-art chatbots internal to the companies.

It is essential to get awareness about what data privacy regulations require of companies when dealing with customer data. The General Data Protection Regulation (GDPR) is the leading regulation from the EU region and it's also relevant to other parts of the world. In the next section, we give a high-level overview of its requirements, which are essential to consider when developing a chatbot.

Introduction to the General Data Protection Regulation (GDPR)

The GDPR is the successor to the 1995 Data Protection Directive, which was a regulation, not a directive. While the directive was left to member states to be transposed into national laws by legislation, the GDPR regulation is immediately enforceable as law in all member states simultaneously. It is a regulation on data protection for European Union citizens. It also applies to the transfer of personal data outside of the EU area. The GDPR gives users control over their personal information and whether they want to share or keep their data private.

It was adopted by all EU states and came into force on May 25, 2018. The regulation enforces hefty fines against non-compliant organizations (fees up to 4% of annual revenues or 20M €, whichever is greater).

Data Protected Under the GDPR

The GDPR in its definition of data is very broad and covers a multiverse of data generated and captured by companies. As per the GDPR, the protected data includes

Necessary identity information (name and surname; date of birth; phone number; a home address; an email address; ID card number and Social Security number etc.); web data (location, IP address, cookie data); health and genetic data; biometric data (data that identifies a person); racial and ethnic origin; religious beliefs; political opinions.

This includes data that chatbots deal with in the course of conversations.

Data Protection Stakeholders

As per the regulation, any company that collects and processes EU citizens' personal information or that stores personal data of EU residents must comply with the GDPR, regardless of whether the company is present in EU territory or not. This scope means that most global businesses need to be GDPR-compliant.

The regulation defines three stakeholders to the GDPR:

- **Data subject**: A person whose data is being processed by a controller or processor.

- **Data controller**: An individual or company that determines the purpose and conditions of collecting and processing personal data from users.

- **Data processor**: An individual or company that processes personal data for data controllers.

The definition of stakeholders directly impacts how we design our chatbots and ensure the rights of our customers who interact with chatbots. For example, a customer interacting with a chatbot is a data subject, and the bank or insurer or company becomes the data controller. The CRM or database system authorized personal also becomes the data controller. If your chatbot uses Dialogflow for processing the data, then it becomes the data processor.

The details of the law can be read from the source here: `https://eur-lex.europa.eu/eli/reg/2016/679/oj`.

Customer Rights Under the GDPR

It is essential for the chatbot developer team and leadership to understand what rights are enshrined in the GDPR for the customers. The chatbot functionality must abide by them.

The rights under GDPR are stated below for your reference:

#	Right	Data Controller Responsibilities
1	**Right to be Informed**	Be transparent in how much you collect and process personal information and the purpose you intend to use it for. Inform your customer of their rights and how to carry them out.
2	**Right of Access**	Your customers have the right to access their data. You need to enable this either through the business process or technical process.
3	**Right to Rectification**	Your customer has the right to correct information that they believe is inaccurate.
4	**Right to Erasure**	You must provide your customer with the right to be forgotten, provided that your legitimate interest to hold such information does not override theirs.
5	**Right to Restriction of Processing**	Your customer has the right to request that you stop processing their data.
6	**Right to Data Portability**	You need to enable the machine and human readable export of your customers' personal information.
7	**Right to Object**	Your customer has the right to object to you using their data.
8	**Right Regarding Automated Decision Making**	You customer has the right not to be subject to a decision based solely on automated processing, including profiling.

Chatbot Compliance to GDPR

In the above sections, we discussed that the chatbots are no longer the subject of business communication only; chatbot makers must consider them in a data controlling and processing manner. This requires the chatbots to face the strict scrutiny of the GDPR.

Some of the generic and minimum steps that the chatbots makers need to take to be ready for GDPR compliance are listed below. The list is not comprehensive; it is just an indicative list for internal assessment. Please consider a full audit of a chatbot before making it public for general use.

- The chatbots, before starting a conversion, must clearly state what data will be collected in the conversation and must be able to access what data is being collected.

- The chatbot user must be allowed to access, review, download, and erase the data collected by the chatbot.

- The chatbot logs must be securely stored and made accessible to users. Also, you must have the explicit permission of the user before processing the log to train your chatbots.

- A clearly stated privacy policy and contact information for a Data Officer for any concerns.

- The option of talking to a real operator rather than a machine chatbot.

These items are an indicative list of the steps that the chatbot owner needs to take. A full audit may reflect more areas to make sure the chatbot is fully compliant.

Summary

In this chapter, we classified conversations into generic and specific areas. While developing a chatbot for general nature requires a multiverse of data, the specific conversation chatbots require only the corpus of data to have those conversations. We introduced different data sources captured from a developer's understanding of the functionality, data generated from customer interactions across all channels, and also, we discussed the significance of open data. Crowdsourcing of data for generic chatbots was also discussed. The significance and challenges of personal data were discussed with examples, and their impact of the design of chatbot was also explained. The most crucial part of a chatbot's implementation is the impact of regulations when chatbots deal with personal data. We introduced the General Data Protection Regulation (GDPR) which protects the data of EU citizens, not only within the EU but outside as well. A short checklist of customers' rights was provided along with some standard steps to be taken for chatbots to be GDRP compliant. In the next chapter, we will discuss how to design the chatbot and create conversation flows for a 24x7 insurance assistant.

CHAPTER 3

Chatbot Development Essentials

Chatbots need to have features that enable human-like conversations. The goal is to make a chatbot conversation more human and thus better than the menu-driven approach of modern apps. In the previous chapters, we discussed types of chatbots and the regulatory constraints to consider for an in-house-developed chatbot. In this chapter, we will discuss the simplified approach to building the integral components of chatbots. Later sections will introduce conversation flow for a sample request to facilitate the context build-up in a chatbot conversation. The chapter will end with the introduction to the "24x7 Insurance Agent" chatbot, which will be the use case discussed throughout the rest of the book.

Customer Service-Centric Chatbots

The customer service process involves the exchange of private customer information and accessing data sources to fetch current information. The information exchange through the chatbot requires accuracy and the

© Abhishek Singh, Karthik Ramasubramanian, Shrey Shivam 2019
A. Singh et al., *Building an Enterprise Chatbot*,
https://doi.org/10.1007/978-1-4842-5034-1_3

enforcement of strict security and privacy policies. There are three critical considerations in chatbot development to achieve the intended use case from a chatbot:

- Business context

- Policy compliance

- Security, authentication, and authorization

The accuracy of a system in resolving input queries is another consideration for the NLP module.

Business Context

Business context refers to the peculiarities of the business the chatbot is developed for. In general conversations, we can infer meanings to unsaid things or implicit things in a conversation.

The concept of live chat is still relevant in many large companies. The core purpose of live chat was to reduce call center costs and provide another layer of conversation, but it also created another conversation silo and, in many cases, lousy customer experience crept in.

In live chats, the business context was maintained by the chat executive and the customer needed to spell out the context every time. The modern chatbot system needs to identify and sustain settings to construct a minimal conversation flow. Usually, the conversations done through chatbots depend on what is explicitly passed to a chatbot for processing (conversation history), what channel is used to converse (channel), and the medium of the message (text, image, voice, emoji, etc.). The unsaid thing is the context, which needs to be maintained by a successful chatbot.

For example, below is a conversation showing how context is important.

> **Conversation context**: User chatting about the food
> order (2345) he canceled last night.

Case 1: No Context Maintained	**Case 2**: Context of Chat Maintained
User: Why refund not processed for my food order?	**User**: Why refund not processed for my food order?
Chatbot: Please provide your OrderID	**Chatbot**: Your refund is processed and will be updated in your account in 2 days.
User: 2345	
Chatbot: Your refund is processed and will be updated in your account in 2 days.	

In Case 1, the context is not maintained, so every conversation needs all the information to be supplied to the chatbot. In Case 2, the chatbot keeps the context of the conversation from where it started and also tries to identify the query to the same meaning.

Building on this generic example, the chatbot needs to be trained in specific business language, terminologies and operating dialect. If you are developing a chatbot for a bank, you need to understand that "time deposit account," "checking account," and "roll over" refer to specific features at a bank. This is very important, since your user won't want to explain what they are talking about to the chatbot time and again.

The business context is built at the Natural Language Processing (NLP) layer, and the conversation framework manages functionality. This specific training for global business context and language requires looking into past conversations with customers.

Note   Every conversation may have a global context and a dialogue-specific context. For example, talking about life insurance is the global context, and within that, talking about a particular policy is the conversation context.

Policy Compliance

Every conversation can lead to many types of requests to be fulfilled by the chatbot. The crucial questions then becomes, to fulfill a user request, what is the compliant way to dispose of that information? The policies and government regulations decide what is allowed and what is not allowed, and if allowed, how to access that information.

For example, suppose a user wants to update his home address for his insurance. This, from a technology perspective, can be as simple as passing the following information to the chatbot (after authentication):

> **User**: Update my home address to XYZ, sector-8, India
>
> **Chatbot**: Address updated to XYZ, sector-8, India

But there are new questions. Is customer allowed to change his address by just providing this command? How would this happen at an actual bank branch? What are the processes and policies that govern an address change?

A policy and regulation guide is essential to drive the conversation through the chatbot. The business process, as well as statutory compliance, needs to be followed across channels for any request to be persisted in the system. For example, updating the address should not be performed even if the request has been confirmed. It may require asking for address proof, reason, waiting three days, or some other process before updating the address.

The chatbots created for customer service need to be carefully taught the must-follow rules or steps to fulfill a request. These rules can differ based on the business and purpose of the chatbot. These steps need to be implemented as strict "AND" conditions in chatbot logic.

Security, Authentication, and Authorization

Chatbots can be allowed to access private information or remain on another channel to distribute public information conversationally. Authenticating and authorization are two essential layers for securing a conversation. A secure communication channel is vital to make sure the data exchanged between the user and the chatbot is encrypted and transferred over a very secure medium.

Security policies can be enforced by having conversations over the HTTPS protocol, having a firewall, and other industry best practices. The security layer must be taken care of by the network technology and technical architecture to secure all the conversations. The security features can be inherited from the existing framework of the business and need not be designed separately, in most cases.

As seen in Figure 3-1, the user first needs to establish the identity via the authentication process. The authentication process generally involves providing a username and password, and in multi-factor authentication, additional information like an OTP or PIN. The successful authentication establishes the user identity as per system records. The next process is to establish the permissions the user has in the system. Authorization is also called role management in business terminology, and hence the business needs to assign roles/permissions for each authenticated user before deployment of the chatbot.

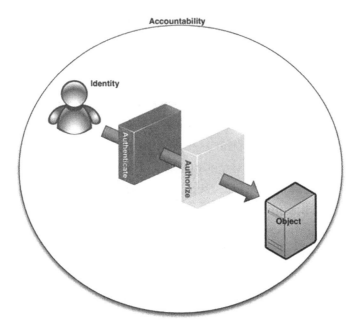

Figure 3-1. *Reference accountability framework*

The chatbot needs to authenticate its users to ascertain if they are allowed to access chatbot features, and if allowed, which features are allowed for which user. This becomes more important when your chatbot connects to the backend system for accessing information like HRM, CRM, and other methods. In many cases, an Identity Management System (IMS) like Active Directory can be used to create authentication and access the control mechanism for the chatbot as well. A typical authentication can be a multi-factor login system or an access PIN for authorized users.

Authorization is another essential layer in the accountability framework and it allows the enterprise to control access to the resources. This also controls the access control layer where the systems check what areas or functionality the authenticated user is allowed to access. This is very important because multiple users will want to use the same source of information in multiple independent conversations, and we want to protect the data and deliver only what the user is authorized to access as per their role and policies.

Figure 3-2 shows how typical access is granted for an application; it's valid for chatbots as well. We can create a full, custom authentication and authorization service or use third-party tools like Auth0, Active Directory, etc. In the shown approach, the application asks the user for their credentials and sends them to the auth service. Once authenticated, it returns a token with authorization details, which can be used by the chatbot to interact with the user.

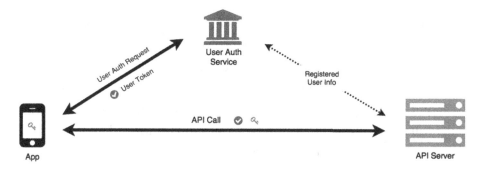

Figure 3-2. *Authentication and authorization service*

Accuracy of User Input Translation to Systems

The chatbot logic creates an interface between the user inputs in natural language and machine actionable inputs for retrieving information. This interface needs to make sure that the translation is accurate before the output is delivered to the user. This is the biggest

41

challenge NLP-based chatbots face today, and hence one of the substantial research areas. Let's explain this with an example:

Input	Chatbot Logic Generated Query	System Output
User: Tell me the status of my salary for April 2019	Select status from payroll_table where empID=UserID and month=" April" and year="2019"	Your salary status is "<RESPONSE FROM QUERY>"
User: When I will get my last month's salary?	Select status from payroll_table where empID=UserID and month= ?? and year= ??	Your salary status is "<RESPONSE FROM QUERY>"

In this example, Input 1 and Input 2 need to get the same result from the system. The chatbot logic needs to resolve for the two inputs before it can fetch the correct information for the user. Here are the challenges and requirements of making sure the query gets accurate data for the system to respond. If the chatbot logic is unable to create the right question, the results will not be correct and may even cause unauthorized information to be shared with users unintentionally.

Chatbot Development Approaches

The chatbot development approach refers to how to build the chatbot logic. The critical consideration in selecting the approach is the balance between natural language abilities and accuracy of results. As you know, NLP comes with a challenge of understanding the natural conversations and translating them to machine actions; a balance must be maintained.

As shown in Figure 3-3, there is a general classification of chatbots based on their abilities and the extent of AI built within them. The critical axis over which we classify chatbots is conversation scope and machine responses.

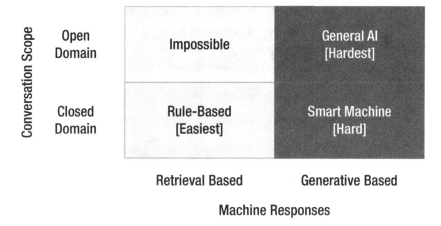

Figure 3-3. *Chatbot classification based on conversation types and response types*

The focus of this book and our discussion is on closed domain conversations, where we can have either a rule-based approach or create smart machines using an AI-based approach. In this section, we describe the two popular types of development approaches.

Rules-Based Approach

The rules-based approach, also called the menu-driven approach, works as an extension of self-help portals with a better experience. The critical difference is with the navigation to solutions. In a self-help portal, you need to navigate to the right options manually, while in menu-based chatbots, the navigation can be done using natural language and then actions are performed using menus. See Figure 3-4.

These kinds of chatbots are prevalent and usually high in usage across industry use cases that are integrated with CRM and other data systems.

Figure 3-4. *Menu-driven chatbot interface*

As shown in Figure 3-4, a chatbot attempts to understand the user question and then presents a menu to choose the next action. The list makes sure the backend knows what exact operation needs to be done to fulfill the request.

Advantages of the Menu-Based Approach

There are some advantages to the menu-based approach:

- The accuracy of the response is confirmed by the design.

- It's based on heuristics rather than complicated NLP, and is easy to understand and implement.

- It's easy to extend to new menu items without retraining the core.

Disadvantages of the Menu-Based Approach

With the advantage of excellent accuracy and easy implementation, there come some limitations:

- The functionality is strictly limited to the templates' code.

- The fulfillment is of two steps: understand the context and bring the menu up. After the menu click, fulfill the request.

- It offers limited natural language conversations since the chatbot do not understand beyond the coded situations.

Even with these limitations, the menu-driven approach is very successful when accuracy is more important than the experience of a natural conversation.

AI-Based Approach

The AI-based approach is based on an advanced NLP engine to support natural language and fulfill the request based on ML algorithms and system integration for dynamic information retrieval. The accuracy of the chatbot is lower at the start and increases over time.

The critical difference between the menu-based and AI-based approach is the NLP engine. An NLP engine is responsible for extracting the information present in the user input. Moreover, based on the extracted information, the chatbot needs to decide the next steps.

As shown in Figure 3-5, the critical role of the NLP engine is to extract information from the natural language input. The accuracy of the information extraction is critical because it will decide the outcome of the conversation and be persisted in the system. The NLP engine needs to extract the information required to instruct the system to act. In the menu-driven approach, the user must engage with menus to select exact details before the system can act.

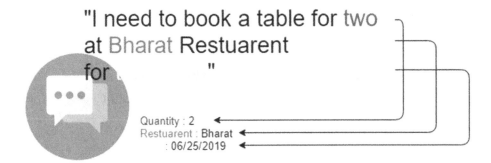

Figure 3-5. *The NLP engine extracts exact information based on ML techniques*

Advantages of the AI-Based Approach

The AI-based approach comes with many advantages and customer-centric benefits:

- Advanced conversations can happen without the user going into multiple steps for actions.

- The NLP engine can deal with unseen scenarios and numerous texts.

- The chatbot can learn to create custom responses from scratch (natural language generation).

Disadvantages of the AI-Based Approach

The problems of the AI-based approach are mostly due to the complexity involved:

- The NLP engine is complex to train, maintain, and improve.

- The accuracy of responses suffers since the NLP output is not 100% correct.

- It requires a vast amount of data for a working chatbot NLP engine.

Conversational Flow

Chatbots for closed-domain applications are built with a defined purpose and functionalities that the chatbot will provide as features to the user. To be able to cover the possible cases of conversation or the user inputs, we must define the scope and all of the flows possible. The flow definitions are essential since we must follow policies to be able to provide access to the required data.

Conversation flow is a decision tree that describes the possible list of events, decisions, and outcomes of a conversation at any point in the conversation. This type of flow secures higher relevance when the context needs to be maintained and the response from the system is not a single step. An example of a conversation flow is shown in Figure 3-6.

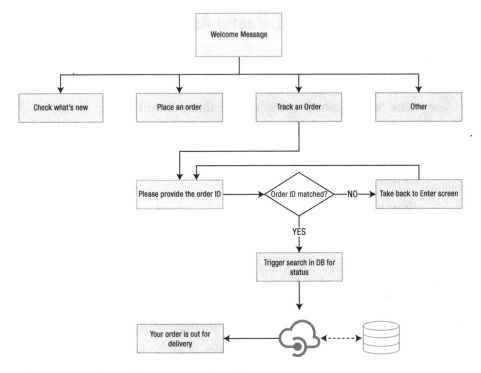

Figure 3-6. *Sample conversation flow*

The flow starts with a welcome message and either provides a menu (if it's a rule-based chatbot) or the user asks a sentence (entirely AI-driven chatbot). Once the chatbot NLP logic identifies which functionality the user requires, there is a decision point to take the user to that conversation path. If the user wants to check the status of an order, his next conversational decision point is to enter the order number. Once the chatbot receives a valid order id, the backend calls for action to retrieve information for that order and take it back to the user. This flow maintains the context as well, so if he needs to track another order, he need not start from the root but can just enter another order id and the chatbot will know to track the order status.

In more advanced chatbots, you can pass multi-intents in one line, but technically the chatbot will process the request in the same flow.

So "Track my order number 465" is a single input from the user, which should fetch the same result. The multi-intent chatbots are difficult to build, and chances of error are high.

Creating a chatbot flow is essential since this defines the scope of features and improves the accuracy of the chatbot for the user. It is imperative to make it clear to the user what the chatbot is meant to do for them and maybe define the features beforehand. The exception cases can always be transferred to default responses or a human executive.

Key Terms in Chatbots

The development of chatbots has become a full-bodied development process, which means it is essential for you to understand the terminology of chatbots before you attempt to develop one. The key terms used in chatbot development also have multiple variants as professed by leading chatbot platform providers like Amazon, Google, etc.

In this section, we will discuss some key terms frequently used in chatbot development. In further chapters, we will use these concepts and terminologies to show how to develop the chatbot from scratch.

Utterance

Utterance refers to anything that user inputs to the chatbot. The full end-to-end input makes an utterance, such as "Get me status of my order id 345," "What is the temperature today?," "Hello," "Good Morning," etc.

Utterances are used to develop a classifier for intents in development. The chatbot stores as many possible utterances in the database, which are the questions asked by the user, and clusters them in different intents, which represent what the user wanted to say.

For developing a chatbot for insurance, we will need to capture actual questions asked by users from different channels, such as chats, emails, office visits, customer value center, etc. We will use all this historical data to train the chatbot to know the actual requirement of the user and which conversational flow to use.

Intent

The intent is the intention of the user identified from the utterance captured by the chatbot. Identifying intent is the essential function of a chatbot. In menu-driven chatbots, the menus help the user to pinpoint the intent, while in AI-based chatbots, identifying intent is a task done by the NLP engine.

The successful matching of intent decides the flow of conversation and delivers the correct response to the user. In domain-specific chatbots, the intents could be different from general intents, and hence domain-specific training is required.

For example, for the utterance of "Show me the stock price of Apple," the intent is to find the stock price. Let's call this intent as showStockPrice. showStockPrice is the main intent of the user, while the term "Apple" is the entity, also called a slot.

Entity

Entity gives meaning to the intent by providing additional value to the utterance. An entity can be defined as subordinate to the intent, which tells us the intent is related to which subclass. In this example, "Apple" is the entity of @company_name for the intent #showStockPrice.

Entities, or slots, when maintained in sessions, help retain the context of the conversation. In this example, after this first utterance we replied with the price. Just after that, the next utterance could be "and for Microsoft." In this case, the chatbot has already captured the intent as showStockPrice, so the slot changes to Microsoft and the chatbot can fetch a stock price of Microsoft.

Channel

The channel is the medium used by the chatbot to connect with the user and fulfill their request. Nowadays, all social media messengers allow chatbots for conversations (e.g., Facebook Messenger, Slack, Skype, etc.).

However, for an application like our 24x7 Insurance Agent, we want to have our developed channel remain compliant with privacy laws and provide an added layer of security when we access the private information of users.

Human Takeover

The human takeover is a term used to denote human fallback during a conversation. Modern chatbots come with the feature of fallback to human assistance when the chatbot fails to understand intents and extract entities, or the confidence is NLP output is low.

Human takeovers can be of two types:

- **A human takeover by choice**: At any moment the user can choose to talk to a human, maybe because they are more comfortable with humans or the chatbot isn't solving their issues.

- **A human takeover by confidence**: A confidence filter may decide if we can fulfill the request with high confidence; if not, the request is automatically transferred to a human without a choice for the user. This provides a seamless experience for the user.

Use Case: 24x7 Insurance Agent

The chatbot that will be discussed throughout the book is based on the use case of an insurance agent. AI-driven chatbots will be discussed using the multiple aspects of security, natural language ML techniques, deployment, and business purposes. Each chapter will discuss some component of developing this chatbot.

Now you know the essential conditions for planning the 24x7 Insurance Agent, including setting the business context, type of development policies, and other considerations. In this section, we will define the aspects for 24x7 Insurance Agent, due to scope all features will not be explicitly implemented.

- **Business context**: The 24x7 Insurance Agent will be able to have necessary conversations in the insurance domain. People may ask about their policies, premiums, etc.

- **Policy compliance**: The policies can be taken from the standard procedure followed at the customer care center.

- **Security, authentication, and authorization**: We can create a PIN-based authorization, while an IDS-based authentication can also be used if the users already have a product.

- **Accuracy of user input translation to systems**: To ensure this, we can create a confidence filter-based human takeover mechanism.

- **The AI-based approach or menu-based approach**: Both should be fine as per need and balance between flexibility and accuracy.

- **Conversational flows**: The flows must be created by sitting with the business and exploring its policies.

- The NLP training will require data of old conversations, intent list, most frequent entities, etc.

These decisions will help the developer to select the right structure and architecture for the chatbot solution.

Summary

Chatbots offer opportunities and challenges. The opportunities outweigh the challenges faced while developing one. In this chapter, we discussed important considerations as you embark on the journey of developing a chatbot, including defining business context, understanding policies to access data, adopting best practices for security of conversations and systems, and ensuring the accuracy of responses. Then we introduced the types of development approaches and their advantages and disadvantages: menu-based and AI-based. The concept of designing conversational flows was covered, which helps create scope and deterministic structure to respond to users. Critical terms used frequently in chatbot development were explained: utterances, intent, entity, channel, and human takeover. In the end, we outlined how to structure the solution for the 24x7 Insurance Agent. The next chapter will introduce the solution architecture and how enterprises can build a successful chatbot in-house.

CHAPTER 4

Building a Chatbot Solution

Chatbots are complete solutions and are created as an independent layer in any solution. The senior management also looks at chatbot functionalities and ROI as an independent entity. The focus on conversational technologies further demands a holistic view on chatbots from solution and business returns perspectives. In previous chapters, we demystified the essentials of developing a chatbot for a closed domain. In this chapter, we will focus on how to build solutions using the best available resources for a closed domain use case. The chapter will also cover a thought process on how to measure success for a chatbot implementation and managing the risks associated with chatbots.

Business Considerations

Any business will ask the question, "What business value does a chatbot add?" This question is to be answered with objectivity and time targets. Technological advancements may allow us to implement advanced chatbots and other solutions, but how they add value to the business is a very subjective call. The business needs to evaluate all factors to ascertain how a chatbot is good for their business.

© Abhishek Singh, Karthik Ramasubramanian, Shrey Shivam 2019
A. Singh et al., *Building an Enterprise Chatbot*,
https://doi.org/10.1007/978-1-4842-5034-1_4

Chatbots vs. Apps

From technological point of view, the business must tackle an important question, specifically relevant to closed domain chatbots: whether to go for an app or a chatbot. In terms of functional features, both can provide the same information for a given feature set. The key differential happens to be chatbots being conversational in nature, while apps are self-service applications.

The key considerations of chatbots vs. apps as mentioned by 2018 State of Chatbot Report (`www.drift.com/wp-content/uploads/2018/01/2018-state-of-chatbots-report.pdf`) are

- Chatbots are preferred to get quick answers for questions and 24-hour access.

- Apps are preferred for ease of use and convenience.

A survey in the report lists some factors that a business must check with their current needs:

- Quick answers to simple questions

- Getting 24-hour service

- Convenience

- Quick answers to complex questions

- Ease of communication

- Ability to easily register a complaint

- Getting detailed/expert answers

- A good customer experience

- Friendliness and approachability

- Having a complaint resolved quickly

Growth of Messenger Applications

Another factor driving the need for chatbots is the increasing usage of messenger applications and a stable growth rate of having a mobile-focused approach by companies. The customer is now connected 24 hours to the Internet through their mobile handsets and wants to access services via easy-to-use interfaces.

In early 2011, messenger applications started coming up and had good adoption rates as mobile devices, internet connectivity, and cloud computing also picked up at same time (www.businessinsider.com/the-messaging-app-report-2015-11). Sometime in early 2015 we got to the point where messenger applications, specifically WhatsApp and Facebook Messenger, were at meteoric adoption rates and equalled the activity on social networks. As the trend suggests, people are more active on messenger applications compared to social media networks.

This trend points out the shift in consumer behaviors where chat is a preferred mode of communication. And this implies that businesses need to enable this channel of communication with customers as well using either chatbots or human chats.

Direct Contact vs. Chat

The increasing use of messenger apps has shifted the way customers want to interact with businesses. In the early days of messenger apps, the studies showed an increasing preference for contact via chats.

The survey by BI Intelligence as summarized by *Chatbots Magazine* (https://chatbotsmagazine.com/chatbot-report-2018-global-trends-and-analysis-4d8bbe4d924b) shows that as early as 2016, the mature market customers adopted chats very quickly. The consumers believed they could solve their issues faster over chat than calling a customer care representative. They also felt more comfortable with chats because they could keep them as a record for follow-ups, unlike phone conversations.

Business Benefits of Chatbots

Considering essential business aspects allows a company to decide whether it wants to go ahead and build a chatbot or improve current apps/channels. If the company decides to go ahead with a chatbot solution, it needs to understand the key value creation by chatbots. Chatbots are now accepted by users with different percentages in different industries, with different adoption rates.

In the aforementioned *Chatbots Magazine* questionnaire, which asked how comfortable you are with being assisted by an AI-based chatbot for business communication, the response showed that people are most ready for such conversations for online retail, generic healthcare queries, and telecommunications.

Further, in the 2018 State of Chatbot Report, the top reasons for not preferring chatbots are the need for assistance from a real person, less awareness of chatbots, and possible blocks due to lack of accessibility of channels (i.e., not having a Facebook account or access to a smartphone).

The studies indicate that there is a value in adding this channel if the benefits for the company are high and your customer is comfortable being assisted by AI-based chatbots. The two topmost value creations from chatbots are discussed in following sections.

Cost Savings

Undoubtedly the most important benefit for the company is cost savings on customer service. The cost savings makes the best case for bringing chatbots into the service delivery channel of company. While there are lot of other strategic and growth benefits, the cost needs to justify all the efforts and resources used for chatbot development and maintenance.

A BI Intelligence study (`www.businessinsider.com/intelligence/research-store?&vertical=mobile#!/The-Messaging-Apps-Report/p/56901061`) shows potential annual savings on salaries by augmenting the chatbots across various business functions. The highest expenditure area for salaries is the customer service representatives, where the savings are also highest. The cost reduction is clearly visible and attributable to the use of chatbots across the functions in insurance sales, reporting, sales, and customer service.

Customer Experience

Customer experience is the second most impactful factor for introducing chatbots to a business. The customer experience brings a multitude of values to the business, not just limited to direct sales or savings. The benefits of good customer experience include

- High brand value and recall

- High lifetime values (improved engagement)

- Brand differentiator from competitors

There are many other derived factors due to a happy customer experience. A loyal customer base is a recipe for long-term success.

The *Chatbots Magazine* summary points out features that contribute to a good and unique customer experience derived from chatbots. The key points are listed below for reference:

- 24-hour service

- Getting an instant response

- Answers to simple questions

- Easy communication

- Complaints resolved quickly

- A good customer experience

- Detailed/expert answers

- Answers to complex questions

- Friendliness and approachability

The variety of new features attracts customers and creates unique value for companies.

Success Metrics

Success metrics are important to define at the start of any chatbot development. The metrics work as a compass to direct the solution and the intended benefits of the chatbot. While there are success metrics related to the accuracy of the NLP engine, the intent classifiers, and other technical aspects, in this section we will only talk about success metrics from a business perspective.

The success metrics need to be manageable and measurable with a simple explanation to the business. We will discuss a few metrics that can be used to track and manage the success of chatbots. The metrics focus on success when you compare a chatbot interaction with a human interaction.

Customer Satisfaction Index

The Customer Satisfaction Index (CSI) measures a customer service representative's quality of interaction by following up with the customer with a small survey and capturing their experience of the interaction. CSI is one of the most impactful metrics to monitor because it provides not only the satisfaction scale but also the areas of improvement.

Completion Rate

The Completion Rate (CR) is defined as the proportion of interactions with a chatbot that ended as the solution resolved for the customer. This metric tells us how many times the chatbot can complete a conversation and deliver the required responses to the user. A higher completion rate indicates a more efficient chatbot service.

Bounce Rate

Bounce rate (BR) can be defined as how many users move away from chat after typing in one or two inputs to the chatbot. A high bounce rate means the chatbot is not successful in engaging the user and this must reflect in some of the customer feedback.

Along with bounce rate, we also measure the reuse rate (RR), which refers to how many customers come back and use the chatbot again. BR is a perfect metric to identify those people who more tend to use chatbots and target similar customer segments.

Managing Risks in Chatbots Service

New technology channels do bring risks. Customers and companies need to understand the risks involved with using chatbots for any transaction or information exchanged through chatbots. The risk is to be understood, communicated, and mitigated before general customers are allowed to use the chatbot services.

Third-Party Channels

Banks and other financial institutions are very much aware about opening a new channel for users to access financial services. While it adds convenience to the customers, it brings some risks as well. Technology is growing way faster than the risk frameworks we have. By the time a general user is able to figure out the risk with usage or best practices for using chatbots, they may already face a security breach.

Top security risks arise from the communication channels for chatbots because they are external to the bank's security control. For example, a customer interacting through Facebook Messenger is interacting with the bank systems using the Facebook platform, which may have vulnerabilities and is not designed for banking operations, just generic chats among people.

In a 2018 survey undertaken by Synopsys, 36% of respondents indicated that customer-facing web applications remain the top security risk to businesses in the Asia-Pacific. September 2018's admission by Facebook that a security breach had affected more than 50 million accounts came as a timely reminder that even tech giants aren't spared. (Source: `finews.asia`).

These cases require financial institutions to limit functionalities through public channels for chatbot messengers. Developing an end-to-end chatbot experience can reduce this risk as well, but adoption remains a challenge.

Impersonation

Another very prominent risk arises from **impersonation**. Impersonation can result in similar looking chatbots, or humans having conversations using fake windows, hacking social media, and other sources of impersonation. The banks already face a lot of fraud due to criminal ingenuity from fraudsters and spend millions in education for phishing, vishing, and other impersonation attacks.

Two-factor authentications are one possible way to reduce the impersonation attacks by having two-step verification from two different sources. In most cases, hackers are not able to crack both factors of authentication and are less likely to be successful in fraudulent transactions.

Personal Information

Personal information revealed through the chatbot channels is a challenge for banks to manage. It is challenging to control users who may accidentally enter their personal information to get access to a service.

As the chatbots are driven by natural language, the chances of revealing personal details are high.

The chatbots need to make sure that they use as little information as possible. It's better to create a PIN with adequate access control so that the user never needs to disclose personal information; they can just use the PIN. Educating the users is an essential step to make sure the user is aware and alert for any fraud.

Confirmation Check

Confirmation is the most impactful and sometimes last resort to make sure transactions done via chatbots are legitimate. Fraudulent or mistaken transactions are possible using chatbots. As new technology comes into user service, it takes time for the users to understand the right use of the service, and in this process, they might do some illegal transactions as well.

For any transaction that seems to be an anomaly or unexpected, it's always good to call up the customer and ask for confirmation before processing it. This check saves the user and bank from fraud.

Generic Solution Architecture for Private Chatbots

Solution architecture is a practice of defining and describing an architecture of a system delivered in context of a specific solution and as such it may encompass a description of an entire system or only its specific parts. Definition of a solution architecture is typically led by a solution architect.

Source: Wikipedia

In this section, we present a reference solution architecture you can apply, with minor modification, to the ideas presented so far (Figure 4-1).

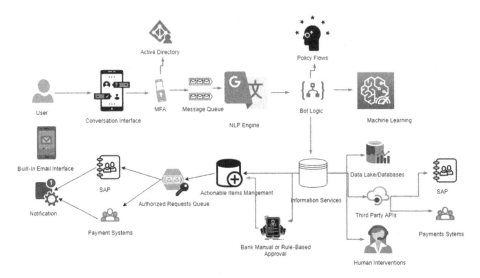

Figure 4-1. *Solution architecture of the 24x7 Insurance Agent*

The architecture captures at a high level how the solution will work. This is not the same as technical architecture, which explains the specific components and their configurations to make the solution work. The precise technical architecture is built as per requirements and is out of scope for the book to cover. The following subsections expand the solution architecture with other vital information to explain it.

Workflow Description

Here is the workflow:

1) **Conversation interface**: We will develop the interface from scratch; it does not depend upon third-party interfaces (e.g., Skype, Telegram, etc.). This will help us create customized interfaces for the organization's need and extend other features as required. The interface can be an entirely new mobile app for the iOS, Android, and Windows platforms.

2) **MFA and Active Directory**: The authentication system will be built at the back end to authenticate devices (MS Intune), users (Active Directory), and applications (by PIN). We will make it on a node environment to allow integration to other identity management services as well. In the natural form, we will only have PIN verification to access the application.

3) **NLP engine**: The NLP engine will be built to accept text inputs from a queue and extract intent and apply context to the incoming query. Once the question is broken into the required components, it's sent to the bot logic.

4) **Bot logic**: This is the core handler of the incoming request. It will have two core inputs before processing the request. The bot logic does not call information services until it has satisfied the process set as per the below two methods. If there isn't enough information for the bot logic to reply, it'll ask the user for more information.

   a) **Policy interaction flows**: These flows are bank expert-designed workflows for the incoming request (e.g., if someone asks for an update of address, what are the essential steps for a reply?) The steps will make sure the user complies with the steps to get an answer. This ensures that all policies, statutory or internal, are followed by the bot, just like an informal HR. Also, policies and FAQ can be defined here.

    b) **Machine learning**: The request that requires a machine learning algorithms to improvise the output is requested from here (e.g., can I request a statement from March 15 to March 18?) This needs a machine to apply appropriate logic to extract dates, an employee id, and an existing account balance to create the right query to the information system. Further, sophisticated features like the mood of the employee, the urgency of the request, and the sentiment of the question will be built here.

5) **Information service**: This is the place where real-time information is fetched for the employee request. This service handles all appeals and prepares responses AND can also send a request for actions as well. The information service will talk to three core data services:

    a) **Data lake/databases**: If some data needs to be fetched from some database or data lake of corporate.

    b) **Third-party APIs/ODBCs**: Interacts with HR systems via ODBC/APIs or some other method that exposes itself with REST APIs.

    c) **Human HR**: If there's a low score of confidence in the reply, it will transfer the request to HR for a response through the chat interface.

6) **Actions**: The information service will also route the requests asking for some work to the actions item management queue (e.g., a request for leave for tomorrow). This request requires an update in PeopleSoft or the HR system. All such change requests will be routed to an HR approval (can be direct as well); once approved, either by real HR or by a policy-based rule, it will be sent to an action queue. HR can support broadcast here, and they can be delivered by an actions workflow.

7) **Updates in real systems**: For authorized and approved requests, we update them directly in the HR system and trigger a notification to the chatbot user and also trigger emails and other built-in process flows of the banking system.

Key Features

Below are the key features:

- **Built for you**: We will make the bot for specific needs, not fitting those needs to existing bot frameworks.

- **Data privacy by design**: The bot is developed with data privacy by design. It will be fully compliant with local laws and internal laws.

- **Developed with a microservices architecture**: The entire application will be based on principles of microservices and hence will allow future-proof design and also advanced application development on top of the framework.

- **Options for deployment**: We can choose which components we want to deploy on-premise or on the public cloud. Based on needs, we can create a deployment plan.

- **Extensible**: We can integrate as many APIs or AI/Ml features as deemed fit for use. All the new future changes can be consumed as APIs in the framework.

Technology Stack

Now the technology stack:

- **Core engine**: Java and JavaScript

- **Backend server**: NodeJS and other JavaScript

- **Front-end server**: Mobile apps based on native frameworks

- **Log management**: Cloud store of a small Hadoop cluster. Also, these stored conversational logs provide data for AI/ML model training.

- **Visualization**: Can be custom created using D3; if self-service is required, then Tableau/PowerBI integration with the logs.

- **Search**: Elastic Search to search the conversation logs.

Maintenance

There are two critical streams in terms of maintenance scope:

- **The technology**: Application uptime will be maintained with the help of on-premise engineers and an on-call channel for all queries.

- **The AI/ML brain**: This will be done offsite by a team of data scientists and updates will be pushed to the systems when ready.

Summary

The chapter described the business considerations for a chatbot solution and listed the benefits of the chatbot. The market growth of messengers is a good sign, indicating that users are adapting to messengers, so a service chatbot can add to their experience and reduce the cost for the company. Third-party studies show the impact chatbots are creating in businesses and how they are bringing digital channels close to natural language. The next section talked about the success metrics that must be defined to manager and improve the chatbot. The essential metrics include the Customer Satisfaction Index (CSI) and Conversion Index (CI).

Further, the chapter also discussed the potential risks of chatbots and how to manage them. The most prominent risks are impersonation and hacking of credentials. Both of these risks require education and secure authentication systems. In the end, we show a reference solution to develop a chatbot. The architecture, workflow, technology stack, and maintenance notes provide enough information to build your chatbot solution as per certain needs. In the following chapters, you will learn about critical areas of natural language sciences, including understanding (NLU), processing (NLP), and generation (NLG). And then we will discuss the implementation of features using open source technology and in-house-developed frameworks.

CHAPTER 5

Natural Language Processing, Understanding, and Generation

The human brain is one of the most advanced machines when it comes to processing, understanding, and generating (P-U-G) natural language. The capabilities of the human brain stretch far beyond just being able to perform P-U-G on one language, dialect, accent, and conversational undertone. No machine has so far reached the human potential of performing all three tasks seamlessly. However, the advances in machine learning algorithms and computing power are making the distant dream of creating human-like bots a possibility.

In this chapter, we will explore the P-U-G of natural languages and their nuances with references to use cases and examples. Table 5-1 provides a quick summary of natural language processing (NLP), natural language understanding (NLU), and natural language generation (NLG) with a few functions and real-world applications. We will get into more details on natural language processing, understanding, and generation in their respective sections.

© Abhishek Singh, Karthik Ramasubramanian, Shrey Shivam 2019
A. Singh et al., *Building an Enterprise Chatbot*,
https://doi.org/10.1007/978-1-4842-5034-1_5

Table 5-1. NLP, NLU, and NLG

Type	NLP	NLU	NLG
Brief	Process and analyze written or spoken text by breaking it down, comprehending its meaning, and determining the appropriate action. It involves parsing, sentence breaking, and stemming.	A specific type of NLP that helps to deal with reading comprehension, which includes the ability to understand meaning from its discourse content and identify the main thought of a passage.	NLG is one of the tasks of NLP to generate natural language text from structured data from a knowledge base. In other words, it transforms data into a written narrative.
Functions	Identify part of speech, text categorizing, named entity recognition, translation, speech recognition	Automatic summarization, semantic parsing, question answering, sentiment analysis	Content determination, document structuring, generating text in interactive conversation
Real-World Application	Article classification for digital news aggregation company	Building a Q&A chatbot, brand sentiment using Twitter and Facebook data	Generating a product description for an e-commerce website or a financial portfolio summary

Chatbot Architecture

When it comes to building an enterprise chatbot, you have so far seen how to identify data sources, design the chatbot architecture, list business use cases, and many other concepts that help an enterprise to process efficiently, reduce manual labor, and reduce the cost of operations. In this chapter, we will focus on the core part of a chatbot: the ability to process textual data and take part in a human-like conversation. Figure 5-1 shows an architecture that utilizes the techniques from NLP, NLU, and NLG to build an enterprise chatbot.

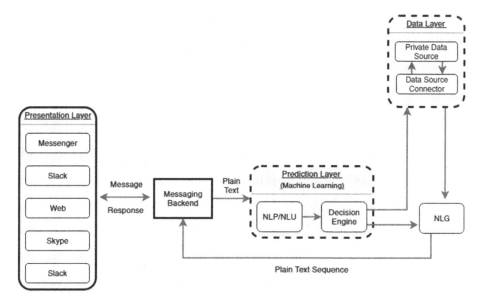

Figure 5-1. *Architecture diagram for chatbots*

Let's say an airline company has built a chatbot to book a flight via their website or social media pages. The following are the steps as per the architecture shown in Figure 5-1:

1. Customer says, "Help me book a flight for tomorrow from London to New York" through the airline's Facebook page. In this case, Facebook becomes the presentation layer. A fully functional chatbot could be integrated into a company's website, social network page, and messaging apps like Skype and Slack.

2. Next, the message is carried to the messaging backend where the plain text passes through an NLP/NLU engine, where the text is broken into tokens, and the message is converted into a machine-understandable command. We will revisit this in greater detail throughout this chapter.

3. The decision engine then matches the command with preconfigured workflows. So, for example, to book a flight, the system needs a source and a destination. This is where NLG helps. The chatbot will ask, *"Sure, I will help in you booking your flight from London to New York. Could you please let me know if you prefer your flight from Heathrow or Gatwick Airport?"* The chatbot picks up the source and destination and automatically generates a follow-up question asking which airport the customer prefers.

4. The chatbot now hits the data layer and fetches
    the flight information from prefed data sources,
    which could typically be connected to live booking
    systems. The data source provides flight availability,
    price, and many other services as per the design.

Some chatbots are heavy on generative responses, and others are built for retrieving information and fitting it in a predesigned conversational flow. For example, in the flight booking use case, we almost know all the possible ways the customer could ask to book a flight, whereas if we take an example of a chatbot for a telemedicine company, we are not sure about all the possible questions a patient could ask. So, in the telemedicine company chatbot, we need the help of generative models built using NLG techniques, whereas in the flight booking chatbot, a good retrieval-based system with NLP and an NLP engine should work.

Since this book is about building an enterprise chatbot, we will focus more on the applications of P-U-G in natural languages rather than going deep into the foundations of the subject. In the next section, we'll show various techniques for NLP and NLU using some of the most popular tools in Python. There are other Java and C# bases libraries; however, Python libraries provide more significant community support and faster development.

Further, to differentiate between NLP and NLU, the Venn diagram in Figure 5-2 shows a few applications of NLP and NLU. It shows NLU as a subset of NLP. The segregation is only in the tasks, not in the scope. The overall objective is to process and understand the natural language text to make machines think like humans.

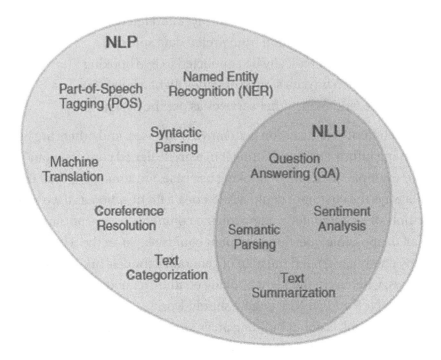

Figure 5-2. *Applications of NLP and NLU*

Popular Open Source NLP and NLU Tools

In this section, we will briefly explore various open source tools available to perform natural language processing, understanding, and generation. While each of these tools does not differentiate between the P-U-G of natural language, we will demonstrate the capabilities of tools under the corresponding three separate headings.

NLTK

The Natural Language Toolkit (NLTK) is a Python library for processing English vocabulary. It has an Apache 2.0 open source license. NLTK is written in the Python programming language. The following are some of the tasks NLTK can perform:

- **Classification of text**: Classifying text into a different category for better organization and content filtering

- **Tokenization of sentences**: Breaking sentences into words for symbolic and statistical natural language processing

- **Stemming words**: Reducing words into base or root form

- **Part-of-speech (POS) tagging**: Tagging the words into POS, which categorizes the words into similar grammatical properties

- **Parsing text**: Determining the syntactic structure of text based on the underlying grammar

- **Semantic reasoning**: Ability to understand the meaning of the word to create representations

NLTK is the first choice of a tool for teaching NLP. It is also widely used as a platform for prototyping and research.

spaCy

Most organizations that build a product involving natural language data are adapting spaCy. It stands out with its offering of a production-grade NLP engine that is accurate and fast. With the extensive documentation, the adaption rate further increases. It is developed in Python and Cython. All the language models in spaCy are trained using deep learning, which provides high accuracy for all NLP tasks.

Currently, the following are some high-level capabilities of spaCy:

- **Covers NLTK features:** Provides all the features of NLTK-like tokenization, POS tagging, dependency trees, named entity recognition, and many more.

- **Deep learning workflow:** spaCy supports deep learning workflows, which can connect to models trained on popular frameworks like Tensorflow, Keras, Scikit-learn, and PyTorch. This makes spaCy the most potent library when it comes to building and deploying sophisticated language models for real-world applications.

- **Multi-language support:** Provides support for more than 50 languages including French, Spanish, and Greek.

- **Processing pipeline:** Offers an easy-to-use and very intuitive processing pipeline for performing a series of NLP tasks in an organized manner. For example, a pipeline for performing POS tagging, parsing the sentence, and named the entity extraction could be defined in a list like this: `pipeline = ["tagger," "parse," "ner"]`. This makes the code easy to read and quick to debug.

- **Visualizers:** Using displaCy, it becomes easy to draw a dependency tree and entity recognizer. We can add our colors to make the visualization aesthetically pleasing and beautiful. It quickly renders in a Jupyter notebook as well.

CoreNLP

Stanford CoreNLP is one of the oldest and most robust tools for all natural language tasks. Its suite of functions offers many linguistic analysis capabilities, including the already discussed POS tagging, dependency tree, named entity recognition, sentiment analysis, and others. Unlike spaCy and NLTK, CoreNLP is written in Java. It also provides Java APIs to use from the command line and third-party APIs for working with modern programming languages. The following are the core features of using CoreNLP:

- **Fast and robust:** Since it is written in Java, which is a time-tested and robust programming language, CoreNLP is a favorite for many developers.

- **A broad range of grammatical analysis:** Like NLTK and spaCy, CoreNLP also provides a good number of analytical capabilities to process and understand natural language.

- **API integration:** CoreNLP has excellent API support for running it from the command line and programming languages like Python via a third-party API or web service.

- **Support multiple Operating Systems (OSs):** CoreNLP works in Windows, Linux, and MacOS.

- **Language support:** Like spaCy, CoreNLP provides useful language support, which includes Arabic, Chinese, and many more.

gensim

gensim is a popular library written in Python and Cython. It is robust and production-ready, which makes it another popular choice for NLP and NLU. It can help analyze the semantic structure of plain-text documents and come out with important topics. The following are some core features of gensim:

- **Topic modeling:** It automatically extracts semantic topics from documents. It provides various statistical models, including latent Dirichlet analysis (LDA) for topic modeling.

- **Pretrained models:** It has many pretrained models that provide out-of-the-box capabilities to develop general-purpose functionalities quickly.

- **Similarity retrieval:** gensim's capability to extract semantic structures from any document makes it an ideal library for similarity queries on numerous topics.

Table 5-2 from the spaCy website summarizes if a given NLP feature is available in NLTK, spaCy, and CoreNLP.

Table 5-2. *Features available in spaCy, NLTK, and CoreNLP*

S.No.	Feature	spaCy	NLTK	CoreNLP
1	Programming language	Python	Python	Java/Python
2	Neural network models	**Yes**	**No**	**Yes**
3	Integrated word vectors	**Yes**	**No**	**No**
4	Multi-language support	**Yes**	**Yes**	**Yes**
5	Tokenization	**Yes**	**Yes**	**Yes**

(*continued*)

Table 5-2. (*continued*)

S.No.	Feature	spaCy	NLTK	CoreNLP
6	Part-of-speech tagging	**Yes**	**Yes**	**Yes**
7	Sentence segmentation	**Yes**	**Yes**	**Yes**
8	Dependency parsing	**Yes**	**No**	**Yes**
9	Entity recognition	**Yes**	**Yes**	**Yes**
10	Entity linking	**No**	**No**	**No**
11	Coreference resolution	**No**	**No**	**Yes**

TextBlob

TextBlob is a relatively less popular but easy-to-use Python library that provides various NLP capabilities like the libraries discussed above. It extends the features provided by NLTK but in a much-simplified form. The following are some of the features of TextBlob:

- **Sentiment analysis:** It provides an easy-to-use method for computing polarity and subjectivity kinds of scores that measures the sentiment of a given text.

- **Language translations:** Its language translation is powered by Google Translate, which provides support for more than 100 languages.

- **Spelling corrections:** It uses a simple spelling correction method demonstrated by Peter Norvig on his blog at http://norvig.com/spell-correct. html. Currently the Engineering Director at Google, his approach is 70% accurate.

fastText

fasText is a specialized library for learning word embeddings and text classification. It was developed by researchers in Facebook's FAI Research (FAIR) lab. It is written in C++ and Python, making it very efficient and fast in processing even a large chunk of data. The following are some of the features of fastText:

- **Word embedding learnings:** Provides many word embedding models using skipgram and Continous Bag of Words (CBOW) by unsupervised training.

- **Word vectors for out-of-vocabulary words:** It provides the capability to obtain word vectors even if the word is not present in the training vocabulary.

- **Text classification:** fastText provides a fast text classifier, which in their paper titled "Bag of Tricks for Efficient Text Classification" claims to be often at par with many deep learning classifiers' accuracy and training time.

In the next few sections, you will see how to apply these tools to perform various tasks in NLP, NLU, and NLG.

Natural Language Processing

Language skills are considered the most sophisticated tasks that a human can perform. Natural language processing deals with understanding and manicuring natural language text or speech to perform specific useful desired tasks. NLP combines ideas and concepts from computer science, linguistics, mathematics, artificial intelligence, machine learning, and psychology.

Mining information from unstructured textual data is not as straightforward as performing a database query using SQL. Categorizing documents based on keywords, identifying a mention of a brand in a social media post, and tracking the popularity of a leader on Twitter are all possible if we can identify entities like a person, organization, and other useful information.

The primary tasks in NLP are processing and analyzing written or spoken text by breaking it down, comprehending its meaning, and determining appropriate action. It involves parsing, sentence breaking, stemming, dependency tree, entity extraction, and text categorization.

We will see how words in a language are broken into smaller tokens and how various transformations work (transforming textual data into a structured and numeric value). We will also explore popular libraries like NLTK, TextBlob, spaCy, CoreNLP, and fastText.

Processing Textual Data

We will use the Amazon Fine Food Review dataset throughout this chapter for all demonstrations using various open-source tools. The dataset can be downloaded from www.kaggle.com/snap/amazon-fine-food-reviews, which is made available with a CC0: Public Domain license.

Reading the CSV File

Using a read_csv function from the pandas library, we read the Reviews. csv file into a food_review data frame and print the top rows (Figure 5-3):

```
import pandas as pd
food_review = pd.read_csv("Reviews.csv")
food_review.head()
```

	Id	ProductId	UserId	ProfileName	HelpfulnessNumerator	HelpfulnessDenominator	Score	Time	Summary	Text
0	1	B001E4KFG0	A3SGXH7AUHU8GW	delmartian	1	1	5	1303862400	Good Quality Dog Food	I have bought several of the Vitality canned d...
1	2	B00813GRG4	A1D87F6ZCVE5NK	dll pa	0	0	1	1346976000	Not as Advertised	Product arrived labeled as Jumbo Salted Peanut...
2	3	B000LQOCH0	ABXLMWJIXXAIN	Natalia Corres "Natalia Corres"	1	1	4	1219017600	"Delight" says it all	This is a confection that has been around a fe...
3	4	B000UA0QIQ	A395BORC6FGVXV	Karl	3	3	2	1307923200	Cough Medicine	If you are looking for the secret ingredient i...
4	5	B006K2ZZ7K	A1UQRSCLF8GW1T	Michael D. Bigham "M. Wassir"	0	0	5	1350777600	Great taffy	Great taffy at a great price. There was a wid...

Figure 5-3. *A CSV file*

As can be seen, the CSV contains columns like ProductID, UserID, Product Rating, Time, Summary, and Text of the review. The file contains almost 500K reviews for various products. Let's sample some reviews to process.

Sampling

Using the sample function from the pandas data frame, let's randomly pick the text of 1000 reviews and print the top rows (see Figure 5-4):

```
food_review_text = pd.DataFrame(food_review["Text"])
food_review_text_1k = food_review_text.sample(n= 1000,random_
state = 123)
food_review_text_1k.head()
```

	Text
277535	I love these chips! They always make a great h...
253901	To add to the pile-on, really really hate the ...
495520	This stuff is the best. I put it on just about...
373115	Organic India Tulsi tea is, to me, the absolut...
547017	I have a German Shorthaired Pointer (3 yrs old...

Figure 5-4. *Samples*

Tokenization Using NLTK

As discussed, NLTK offers many features for processing textual data. The first step in processing text data is to separate a sentence into individual words. This process is called tokenization. We will use the NLTK's word_ tokenize function to create a column in the food_review_text_1k data frame we created above and print the top six rows to see the output of tokenize (Figure 5-5):

```
food_review_text_1k['tokenized_reviews'] = food_review_
text_1k['Text'].apply(nltk.word_tokenize)
food_review_text_1k.head()
```

	Text	tokenized_reviews
277535	I love these chips! They always make a great h...	[I, love, these, chips, !, They, always, make,...
253901	To add to the pile-on, really really hate the ...	[To, add, to, the, pile-on, ,, really, really,...
495520	This stuff is the best. I put it on just about...	[This, stuff, is, the, best, ., I, put, it, on...
373115	Organic India Tulsi tea is, to me, the absolut...	[Organic, India, Tulsi, tea, is, ,, to, me, ,,...
547017	I have a German Shorthaired Pointer (3 yrs old...	[I, have, a, German, Shorthaired, Pointer, (, ...

Figure 5-5. *Top rows*

Word Search Using Regex

Now that we have the tokenized text for each review, let's take the first row in the data frame and search for the presence of the word using a regular expression (regex). The regex searches for any word that contains **c** as its first character and **i** as the third character. We can write various regex searches for a pattern of interest. We use the re.search() function to perform this search:

#Search: All 5-letter words with c as its first letter and i as its third letter

85

```
search_word = set([w for w in food_review_text_1k['tokenized_
reviews'].iloc[0] if re.search('^c.i..$', w)])
print(search_word)
```

```
{'chips'}
```

Word Search Using the Exact Word

Another way of searching for a word is to use the exact word. This can be achieved using the `str.contains()` function in pandas. In the following example, we search for the word "great" in all of the reviews. The rows of the reviews containing the word will be retrieved. They can be considered a positive review. See Figure 5-6.

```
#Search for the word "great" in reviews
food_review_text_1k[food_review_text_1k['Text'].str.contains('great')]
```

	Text	tokenized_reviews
277535	I love these chips! They always make a great h...	[I, love, these, chips, !, They, always, make,...
547017	I have a German Shorthaired Pointer (3 yrs old...	[I, have, a, German, Shorthaired, Pointer, (, ...
153491	Our GreatDane loves these , he's never happy w...	[Our, GreatDane, loves, these, ,, he, 's, neve...
307887	My parents' dog refused to take her medicine u...	[My, parents, ', dog, refused, to, take, her, ...
189614	Like most of the other reviews state, you can ...	[Like, most, of, the, other, reviews, state, ,...
362712	We own two dogs who are drastically different ...	[We, own, two, dogs, who, are, drastically, di...
564730	A friend's daughter has just gone to college. ...	[A, friend, 's, daughter, has, just, gone, to,...
353896	Scents: Cool Impact Arctic Edge - my...	[Scents, :, <, br, /, >, Cool, Impact, <, br, ...
87831	My little diabetic shih-tzu, Lily, is notoriou...	[My, little, diabetic, shih-tzu, ,, Lily, ,, i...
291616	This tastes great on chicken and shrimp and is...	[This, tastes, great, on, chicken, and, shrimp...
48717	This is great tasting Organic Honey. I will d...	[This, is, great, tasting, Organic, Honey, ., ...
40295	this was great....I gave it to my daughter she...	[this, was, great, ..., .I, gave, it, to, my, ...
436373	Not bad...just a little bland and not much in ...	[Not, bad, ..., just, a, little, bland, and, n...
360593	My Mom is a great biscotti baker, she has trad...	[My, Mom, is, a, great, biscotti, baker, ,, sh...

Figure 5-6. Samples with a specific word

NLTK

In this section, we will use many of the features from NLTK for NLP, such as normalization, noun phrase chunking, named entity recognition, and document classifier.

Normalization Using NLTK

In many natural language tasks, we often deal with the root form of the words. For example, for the words "baking" and "baked," the root word is "bake." This process of extracting the root word is called stemming or normalization. NLTK provides two functions implementing the stemming algorithm. The first is the Porter Stemming algorithm, and the second is the Lancaster stemmer.

There are slight differences in the quality of output from both algorithms. For example, in the following example, the Porter stemmer converts the word "sustenance" into "sustain" while the Lancaster stemmer outputs "sust."

```
words = set(food_review_text_1k['tokenized_reviews'].iloc[0])
print(words)

porter = nltk.PorterStemmer()
print([porter.stem(w) for w in words])
```

Before
```
{'when', 'always', 'great', 'vending', 'for', 'make', "'m",
'just', 'I', '.', 'love', 'a', 'They', 'with', 'healthy',
'these', 'snack', 'the', 'at', 'work', 'chips', 'machine',
'stuck', 'sustenance', '!'}
```

After
```
['when', 'alway', 'great', 'vend', 'for', 'make', "'m", 'just',
'I', '.', 'love', 'a', 'they', 'with', 'healthi', 'these',
'snack', 'the', 'at', 'work', 'chip', 'machin', 'stuck',
'susten', '!']
```

```
lancaster = nltk.LancasterStemmer()
print([lancaster.stem(w) for w in words])
```

```
['when', 'alway', 'gre', 'vend', 'for', 'mak', "'m", 'just',
'i', '.', 'lov', 'a', 'they', 'with', 'healthy', 'thes',
'snack', 'the', 'at', 'work', 'chip', 'machin', 'stuck',
'sust', '!']
```

Noun Phrase Chunking Using Regular Expressions

Above you saw the tokens as a fundamental unit in any NLP processing. Since in natural language, a group of tokens combined often reveals the meaning or represents a concept, we create chunks. Multi-token sequences are created by segmenting using a process called chunking. In Figure 5-7, the smaller boxes show word-level tokenization and the larger boxes shows multi-token sequences, also called higher-level chunks. Such chunks are created using regular expressions or by using the n-gram (more on this in later sections) method. Chunking is essential for entity recognition, which we will shortly explore.

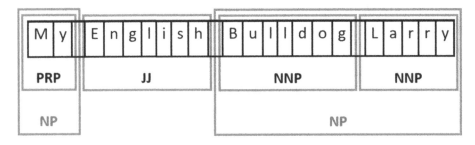

Figure 5-7. *Tokens and chunks*

Let's consider a single review as shown in the following code. The grammar finds a noun using a rule that says, *find noun chunk where zero or one (?) determiner (DT) is followed by any number (*) of adjectives (JJ) and a noun (NN)*. In the POS tree shown in the output of the following code, all the chunks marked as NP are the noun phrases:

```python
import nltk
from nltk.tokenize import word_tokenize

#Noun phrase chunking
text = word_tokenize("My English Bulldog Larry had skin
allergies the summer we got him at age 3, I'm so glad that now
I can buy his food from Amazon")

#This grammar rule: Find NP chunk when an optional determiner
(DT) is followed by any number of adjectives (JJ) and then a
noun (NN)

grammar = "NP: {<DT>?<JJ>*<NN>}"

#Regular expression parser using the above grammar
cp = nltk.RegexpParser(grammar)

#Parsed text with pos tag
review_chunking_out = cp.parse(nltk.pos_tag(text))

#Print the parsed text
print(review_chunking_out)

(S
  My/PRP$
  English/JJ
  Bulldog/NNP
  Larry/NNP
  had/VBD
  skin/VBN
```

```
allergies/NNS
(NP the/DT summer/NN)
we/PRP
got/VBD
him/PRP
at/IN
(NP age/NN)
3/CD
,/,
I/PRP
'm/VBP
so/RB
glad/JJ
that/IN
now/RB
I/PRP
can/MD
buy/VB
his/PRP$
(NP food/NN)
from/IN
Amazon/NNP)
```

You can see many NPs such as "the summer" and "age" where "the summer" is not a single word token. Above you see that the POS is in a tree representation. Another way of representing the chunk structures is by using tags. The IOB tag representation is a general standard. In this scheme, each token is represented as I (Inside), O (Outside), and B (Begin). Chunk tag B represents the beginning of a chunk. Subsequent tokens within a chunk are tagged I and all other tokens are tagged O. Figure 5-8 provides one example of an IOB tag representation.

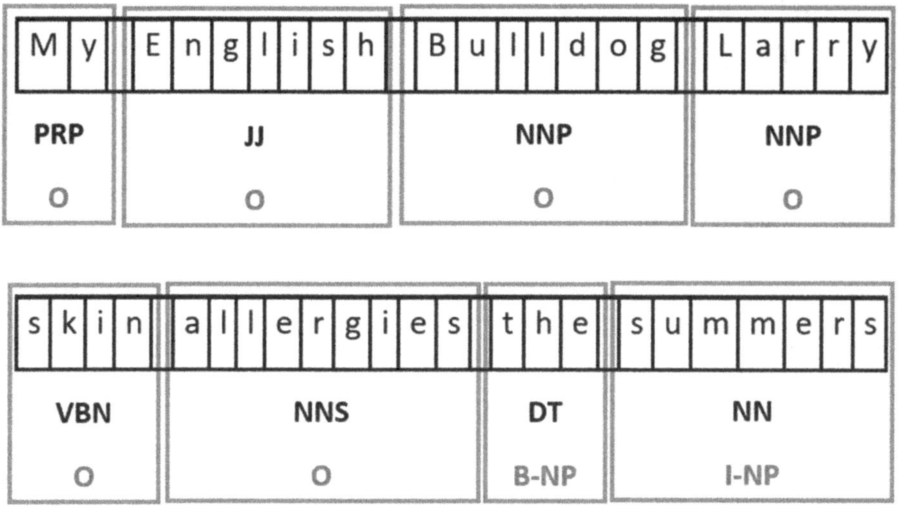

M	y		E	n	g	l	i	s	h		B	u	l	l	d	o	g		L	a	r	r	y
PRP			**JJ**								**NNP**								**NNP**				
O			O								O								O				

s	k	i	n		a	l	l	e	r	g	i	e	s		t	h	e		s	u	m	m	e	r	s
VBN					**NNS**										**DT**				**NN**						
O					O										B-NP				I-NP						

Figure 5-8. *IOB tag representation of chunk structures*

The following code uses the CoNLL 2000 Corpus to convert the tree to tags using the function tree2conlltags(). CoNLL is *Wall Street Journal* text that has been tagged and chunked using IOB notation.

```
from nltk.chunk import conlltags2tree, tree2conlltags
from pprint import pprint

#Print IOB tags
review_chunking_out_IOB = tree2conlltags(review_chunking_out)
pprint(review_chunking_out_IOB)

[('My', 'PRP$', 'O'),
 ('English', 'JJ', 'O'),
 ('Bulldog', 'NNP', 'O'),
 ('Larry', 'NNP', 'O'),
 ('had', 'VBD', 'O'),
 ('skin', 'VBN', 'O'),
 ('allergies', 'NNS', 'O'),
 ('the', 'DT', 'B-NP'),
```

```
('summer', 'NN', 'I-NP'),
('we', 'PRP', 'O'),
('got', 'VBD', 'O'),
('him', 'PRP', 'O'),
('at', 'IN', 'O'),
('age', 'NN', 'B-NP'),
('3', 'CD', 'O'),
(',', ',', 'O'),
('I', 'PRP', 'O'),
('"m', 'VBP', 'O'),
('so', 'RB', 'O'),
('glad', 'JJ', 'O'),
('that', 'IN', 'O'),
('now', 'RB', 'O'),
('I', 'PRP', 'O'),
('can', 'MD', 'O'),
('buy', 'VB', 'O'),
('his', 'PRP$', 'O'),
('food', 'NN', 'B-NP'),
('from', 'IN', 'O'),
('Amazon', 'NNP', 'O')]
```

Named Entity Recognition

Once we have the POS of the text, we can extract the named entities.
Named entities are definite noun phrases that refer to specific
individuals such as ORGANIZATION and PERSON. Some other entities
are LOCATION, DATE, TIME, MONEY, PERCENT, FACILITY, and
GPE. The facility is any human-made artifact in the architecture and civil
engineering domain, such as Taj Mahal or Empire State Building. GPE
means geopolitical entities such as city, state, and country. We can extract
all these entities using the ne_chunk() method in the nltk library.

The following code uses the POS tagged sentence and applies the ne_chunk() method to it. It identifies *Amazon* as GPE and *Bulldog Larry* as a PERSON. In this case, this is both true and false. Amazon is identified as ORGANIZATION, which we expect here. Later in the chapter, we will train our own named entity recognizer to improve the performance.

```
tagged_review_sent = nltk.pos_tag(text)
print(nltk.ne_chunk(tagged_review_sent))

(S
  My/PRP$
  English/JJ
  (PERSON Bulldog/NNP Larry/NNP)
  had/VBD
  skin/VBN
  allergies/NNS
  the/DT
  summer/NN
  we/PRP
  got/VBD
  him/PRP
  at/IN
  age/NN
  3/CD
  ,/,
  I/PRP
  'm/VBP
  so/RB
  glad/JJ
  that/IN
  now/RB
  I/PRP
  can/MD
```

```
buy/VB
his/PRP$
food/NN
from/IN
(GPE Amazon/NNP))
```

spaCy

While spaCy offers all the features of NLTK, it is regarded as one of the best production grade tools for an NLP task. In this section, we will see how to use the various methods provided by the spaCy library in Python.

spaCy provides three core models: en_core_web_sm (10MB), en_core_web_md (91MB), and en_core_web_lg (788MB). The larger model is trained on bigger vocabulary and hence will give higher accuracy. So depending on your use case, choose wisely the model that fits your requirements.

POS Tagging

After loading the model using `spaCy.load()`, you can pass any string to the model, and it provides all the methods in one go. To extract POS, the `pos_method` is used. In the following code, after tokenizing, we print the following:

- `text`: The original text

- `lemma`: Token after stemming, which is the base form of the word

- `pos`: Part of speech

- `tag`: POS with details

- `dep`: The relationship between the tokens. Also called syntactical dependency.

- shape: The shape of the word (i.e., capitalization, punctuation, digits)

- is_alpha: Returns True if the token is an alphanumeric character

- is.stop: Returns True if the token is a stopword like "at," "so," etc.

```
# POS tagging
import spacy

nlp = spacy.load('en_core_web_sm')
doc = nlp(u"My English Bulldog Larry had skin allergies the
summer we got him at age 3, I'm so glad that now I can buy his
food from Amazon")

for token in doc:
    print(token.text, token.lemma_, token.pos_, token.tag_,
    token.dep_, token.shape_, token.is_alpha, token.is_stop)

My -PRON- DET PRP$ poss Xx True True
English English PROPN NNP compound Xxxxx True False
Bulldog Bulldog PROPN NNP nsubj Xxxxx True False
Larry Larry PROPN NNP nsubj Xxxxx True False
had have VERB VBD ccomp xxx True True
skin skin NOUN NN compound xxxx True False
allergies allergy NOUN NNS dobj xxxx True False
the the DET DT det xxx True True
summer summer NOUN NN npadvmod xxxx True False
we -PRON- PRON PRP nsubj xx True True
got get VERB VBD relcl xxx True False
him -PRON- PRON PRP dobj xxx True True
```

```
at at ADP IN prep xx True True
age age NOUN NN pobj xxx True False
3 3 NUM CD nummod d False False
, , PUNCT , punct , False False
I -PRON- PRON PRP nsubj X True True
'm be VERB VBP ROOT 'x False True
so so ADV RB advmod xx True True
glad glad ADJ JJ acomp xxxx True False
that that ADP IN mark xxxx True True
now now ADV RB advmod xxx True True
I -PRON- PRON PRP nsubj X True True
can can VERB MD aux xxx True True
buy buy VERB VB ccomp xxx True False
his -PRON- DET PRP$ poss xxx True True
food food NOUN NN dobj xxxx True False
from from ADP IN prep xxxx True True
Amazon Amazon PROPN NNP pobj Xxxxx True False
```

Dependency Parsing

The spaCy dependency parser has a rich API which helps to navigate the dependency tree. It also provides the capability to detect sentence boundaries and iterate through the noun phrase or chunks. In the following example, the noun_chunks method in the model is iteratable with the following methods:

- text: Original noun chunk

- root.text: Original word connecting the noun chunk to the rest of the noun chunk parse

- root.dep: Dependency relation connecting the root to its head

- root.head: Root token's head

From the example, **"My English Bulldog"** is a noun phrase, where "**Bulldog**" is root text with **"nsubj"** relation and **"had"** as its root head.

#Dependency parse

```
import spacy

nlp = spacy.load("en_core_web_sm")
doc = nlp(u"My English Bulldog Larry had skin allergies the
summer we got him at age 3, I'm so glad that now I can buy his
food from Amazon")
for chunk in doc.noun_chunks:
    print(chunk.text, chunk.root.text, chunk.root.dep_,
            chunk.root.head.text)

My English Bulldog Bulldog nsubj had
Larry Larry nsubj had
skin allergies allergies dobj had
we we nsubj got
him him dobj got
age age pobj at
I I nsubj 'm
I I nsubj buy
his food food dobj buy
Amazon Amazon pobj from
```

Dependency Tree

spaCy provides a method called displayCy for visualization. We can draw the dependency tree of a given sentence using displaCy (see Figures 5-9, 5-10, and 5-11).

```
import spacy
from spacy import displacy

nlp = spacy.load("en_core_web_sm")
doc = nlp(u"My English Bulldog Larry had skin allergies the
summer we got him at age 3")

displacy.render(doc, style='dep')
```

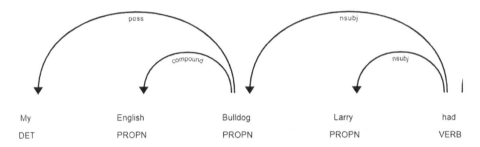

Figure 5-9. *Dependency tree, part 1*

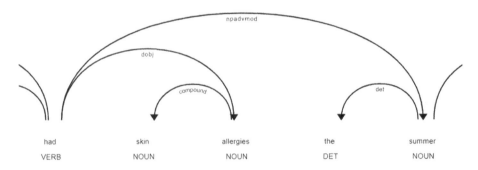

Figure 5-10. *Dependency tree, part 2*

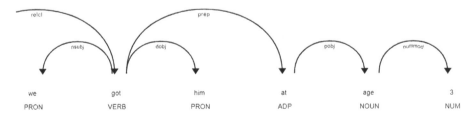

Figure 5-11. *Dependency tree, part 3*

From the dependency trees, you can see that there are two compound word pairs, "English Bulldog" and "skin allergies," and NUM **"3"** is the modifier of **"age."** You can also see "summer" as the noun phrase as an adverbial modifier (npadvmod) to the token "had." You can also observe many direct objects (dobj) of a verb phrase, which is a noun phrase, like (got, him) and (had, allergies) and object of a preposition (pobj) like (at, age). A detailed explanation of the relationships in a dependency tree can be found here: https://nlp.stanford.edu/software/dependencies_manual.pdf.

Chunking

spaCy provides an easy-to-use retrieval of chunk information such as VERB and NOUN from a given text. The noun_chunks method provides noun phrases, and from pos, we can search for VERB. The following code extracts noun phrases and verbs from the chunks:

```
# pip install spacy
# python -m spacy download en_core_web_sm

import spacy

# Load English tokenizer, tagger, parser, NER, and word vectors
nlp = spacy.load("en_core_web_sm")

# Process whole documents
text = ("My English Bulldog Larry had skin allergies the summer
we got him at age 3, I'm so glad that now I can buy his food
from Amazon")
doc = nlp(text)

# Analyze syntax
print("Noun phrases:", [chunk.text for chunk in doc.noun_chunks])
print("Verbs:", [token.lemma_ for token in doc if token.pos_ ==
"VERB"])
```

```
Noun phrases: ['My English Bulldog', 'Larry', 'skin allergies',
'we', 'him', 'age', 'I', 'I', 'his food', 'Amazon']
Verbs: ['have', 'get', 'be', 'can', 'buy']
```

Named Entity Recognition

spaCy has an accuracy of 85.85% in named entity recognition (NER) tasks. The en_core_web_sm model provides the function *ents,* which provides the entities. The model is trained on the OntoNotes dataset, which can be found at https://catalog.ldc.upenn.edu/LDC2013T19.

The default models in spaCy provide the entities shown in Table 5-3.

Table 5-3. *Types*

TYPE	DESCRIPTION
PERSON	Names of people including fictional characters
NORP	Nationalities or religious or political groups
FAC	Civil engineering structures or infrastructures like buildings, airports, highways, bridges, etc.
ORG	Organization names like companies, agencies, institutions, etc.
GPE	A geopolitical entity like countries, cities, states
LOC	Non-GPE locations like mountain ranges, water bodies
PRODUCT	Objects, vehicles, foods, etc. (not services)
EVENT	Named hurricanes, battles, wars, sports events, etc.
WORK_OF_ART	Titles of books, songs, etc.
LAW	Named documents made into laws
LANGUAGE	Any named language

(continued)

Table 5-3. (*continued*)

TYPE	DESCRIPTION
DATE	Absolute or relative dates or periods
TIME	Times smaller than a day
PERCENT	Percentage, including %
MONEY	Monetary values, including unit
QUANTITY	Measurements, as of weight or distance
ORDINAL	"first," "second," etc.
CARDINAL	Numerals that do not fall under another type

The following code extracts the English Bulldog Larry as a PERSON entity and Amazon as entity ORG. Unlike NLTK, where it identified Amazon as GPE, spaCy correctly identifies the context of the sentence to figure out that Amazon is an organization in the given sentence.

```
import spacy

# Load English tokenizer, tagger, parser, NER, and word vectors
nlp = spacy.load("en_core_web_sm")

# Process whole documents
text = ("My English Bulldog Larry had skin allergies the summer
we got him at age 3, I'm so glad that now I can buy his food
from Amazon")
doc = nlp(text)

# Find named entities
for entity in doc.ents:
    print(entity.text, entity.label_)

English Bulldog Larry PERSON
Amazon ORG
```

We can also visualize the entities using the `display` method (shown in Figure 5-12):

```
import spacy
from spacy import display
from pathlib import Path

text = "I found these crisps at our local WalMart & figured I
would give them a try. They were so yummy I may never go back
to regular chips, not that I was a big chip fan anyway. The
only problem is I can eat the entire bag in one sitting. I give
these crisps a big thumbs up!"

nlp = spacy.load("en_core_web_sm")
doc = nlp(text)
svg = displacy.serve(doc, style="ent")
output_path = Path("images/sentence_ne.svg")
output_path.open("w", encoding="utf-8").write(svg)
```

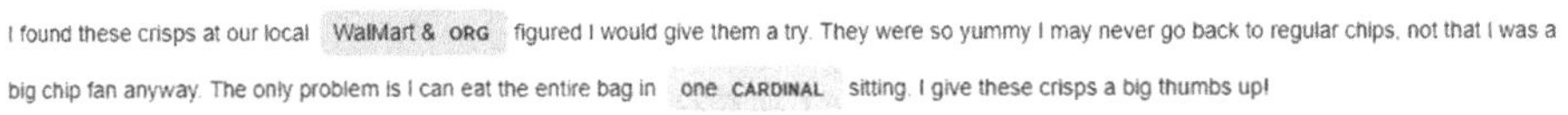

I found these crisps at our local   WalMart & ORG   figured I would give them a try. They were so yummy I may never go back to regular chips, not that I was a big chip fan anyway. The only problem is I can eat the entire bag in   one CARDINAL   sitting. I give these crisps a big thumbs up!

Figure 5-12. *Results*

Pattern-Based Search

spaCy also provides a pattern or rule-based search. We can define our pattern on top of a function like LOWER. For example, in the following code, we define a search span as "Walmart" in lowercase followed by a punctuation mark. This pattern could be written like

```
pattern = [{"LOWER": "walmart"}, {"IS_PUNCT": True}]
```

In the search span, if we want to find the word "Walmart," we define this using the `matcher.add` method and pass `pattern` as the argument to the method.

This syntax is more user-friendly than a cumbersome regular expression, which is hard to understand. The result of the search reveals that the word "Walmart" is found at the seventh position in the string and ends at the ninth position. The output also shows the span text as "Walmart &," which we defined in the pattern.

```
# Spacy - Rule-based matching

import spacy
from spacy.matcher import Matcher

nlp = spacy.load("en_core_web_sm")
matcher = Matcher(nlp.vocab)
#Search for Walmart after converting the text in lower case and
pattern = [{"LOWER": "walmart"}, {"IS_PUNCT": True}]
matcher.add("Walmart", None, pattern)

doc = nlp(u"I found these crisps at our local WalMart & figured
I would give them a try. They were so yummy I may never go back
to regular chips, not that I was a big chip fan anyway. The
only problem is I can eat the entire bag in one sitting. I give
these crisps a big thumbs up!")

matches = matcher(doc)

for match_id, start, end in matches:
    string_id = nlp.vocab.strings[match_id]  # Get string
    representation
    span = doc[start:end]  # The matched span
    print(match_id, string_id, start, end, span.text)

16399848736434528297 Walmart 7 9 WalMart &
```

Searching for Entity

Using the matcher method, we can also search for a type of entity in a given text. In the following code, we search for the entity ORG (defined by "label") named "walmart."

```
from spacy.lang.en import English
from spacy.pipeline import EntityRuler

nlp = English()
ruler = EntityRuler(nlp)
patterns = [{"label": "ORG","pattern":[{"lower":"walmart"}]}]

ruler.add_patterns(patterns)
nlp.add_pipe(ruler)

doc = nlp(u"I found these crisps at our local WalMart & figured
I would give them a try. They were so yummy I may never go back
to regular chips, not that I was a big chip fan anyway. The
only problem is I can eat the entire bag in one sitting. I give
these crisps a big thumbs up!")
print([(ent.text, ent.label_) for ent in doc.ents])

[('WalMart', 'ORG')]
```

Training a Custom NLP Model

In many real-world datasets, the entities are not detected as per the expectations. The models in spaCy or NLTK are not trained on those words or tokens. In such cases, we can train a custom model using a private dataset. We have to create training data in a particular format. In the following code, we pick two sentences and tag an entity PRODUCT with its start and end position in the text. The syntax looks like this:

```
(
    u"As soon as I tasted one and it tasted like a corn chip I
    checked the ingredients. ",
    {
    "entities": [(45, 49, "PRODUCT")]
    }
)
```

We tag the food product "corn" in the two sentences. Here we take just two sentences, and spaCy trains the model well with just them. If you don't get the right entity with a smaller dataset, you might need to add a few more examples before the model will pick the right entity.

```
import spacy
import random

train_data = [
        (u"As soon as I tasted one and it tasted like a corn
        chip I checked the ingredients. ", {"entities": [(45,
        49, "PRODUCT")]}),
        (u"I found these crisps at our local WalMart & figured
        I would give them a try", {"entities": [(14, 20,
        "PRODUCT")]})
]

other_pipes = [pipe for pipe in nlp.pipe_names if pipe != "ner"]

with nlp.disable_pipes(*other_pipes):
    optimizer = nlp.begin_training()
    for i in range(10):
        random.shuffle(train_data)
```

```
for text, annotations in train_data:
        nlp.update([text], [annotations], sgd=optimizer)
nlp.to_disk("model/food_model")
```

We saved the trained model to disk and named it food_model. In the following code, we load the food_model from disk and try to predict the entity on a test sentence. We see it does a good job here in identifying corn as a PRODUCT entity.

```
import spacy
nlp = spacy.load("model/food_model")
text = nlp("I consume about a jar every two weeks of this,
either adding it to fajitas or using it as a corn chip dip")

for entity in text.ents:
    print(entity.text, entity.label_)
```

```
corn PRODUCT
```

CoreNLP

CoreNLP is another popular toolkit for linguistic analysis such as POS tagging, dependency tree, named entity recognition, sentiment analysis, and many others. We are going to use the CoreNLP features from Python through a third-party wrapper called Stanford-corenlp. It can be installed using pip install in the command line or cloned from GitHub here: https://github.com/Lynten/stanford-corenlp.

Once you install or download the code, you need to specify the path to the Stanford-corenlp code from where it picks up the necessary model for the various NLP tasks.

Tokenizing

As with NLTK and spaCy, you can extract words or tokens. The model provides a method named word_tokenize for performing the tokenization:

```
# Simple usage
from stanfordcorenlp import StanfordCoreNLP

nlp = StanfordCoreNLP(<Path to CoreNLP folder>)

sentence = 'I consume about a jar every two weeks of this,
either adding it to fajitas or using it as a corn chip dip'

print('Tokenize:', nlp.word_tokenize(sentence))

Tokenize: ['I', 'consume', 'about', 'a', 'jar', 'every', 'two',
'weeks', 'of', 'this', ',', 'either', 'adding', 'it', 'to',
'fajitas', 'or', 'using', 'it', 'as', 'a', 'corn', 'chip', 'dip']
```

Part-of-Speech Tagging

POS can be extracted using the method pos_tag in the stanford-corenlp:

```
print('Part of Speech:', nlp.pos_tag(sentence))

Part of Speech: [('I', 'PRP'), ('consume', 'VBP'), ('about',
'IN'), ('a', 'DT'), ('jar', 'NN'), ('every', 'DT'), ('two',
'CD'), ('weeks', 'NNS'), ('of', 'IN'), ('this', 'DT'), (',',
','), ('either', 'CC'), ('adding', 'VBG'), ('it', 'PRP'),
('to', 'TO'), ('fajitas', 'NNS'), ('or', 'CC'), ('using',
'VBG'), ('it', 'PRP'), ('as', 'IN'), ('a', 'DT'), ('corn',
'NN'), ('chip', 'NN'), ('dip', 'NN')]
```

Named Entity Recognition

Stanford-corenlp provides the method ner to extract the named entities. Observe that the output by default is in the IOB (Inside, Outside, and Begin) format.

```
print('Named Entities:', nlp.ner(sentence))
```

```
Named Entities: [('I', 'O'), ('consume', 'O'), ('about', 'O'),
('a', 'O'), ('jar', 'O'), ('every', 'SET'), ('two', 'SET'),
('weeks', 'SET'), ('of', 'O'), ('this', 'O'), (',', 'O'),
('either', 'O'), ('adding', 'O'), ('it', 'O'), ('to', 'O'),
('fajitas', 'O'), ('or', 'O'), ('using', 'O'), ('it', 'O'),
('as', 'O'), ('a', 'O'), ('corn', 'O'), ('chip', 'O'), ('dip',
'O')]
```

Constituency Parsing

Constituency parsing extracts a constituency-based parse tree from a given sentence that is representative of the syntactic structure according to a phase structure grammar. See Figure 5-13 for a simple example.

```
print('Constituency Parsing:', nlp.parse(sentence))
```

```
Constituency Parsing: (ROOT
  (S
    (NP (PRP I))
    (VP (VBP consume)
      (PP (IN about)
        (NP (DT a) (NN jar)))
      (NP
        (NP (DT every) (CD two) (NNS weeks))
        (PP (IN of)
          (NP (DT this))))
```

```
(, ,)
(S
  (VP (CC either)
    (VP (VBG adding)
      (NP (PRP it))
      (PP (TO to)
        (NP (NNS fajitas))))       .
    (CC or)
    (VP (VBG using)
      (NP (PRP it))
      (PP (IN as)
        (NP (DT a) (NN corn) (NN chip) (NN dip)))))))))))
```

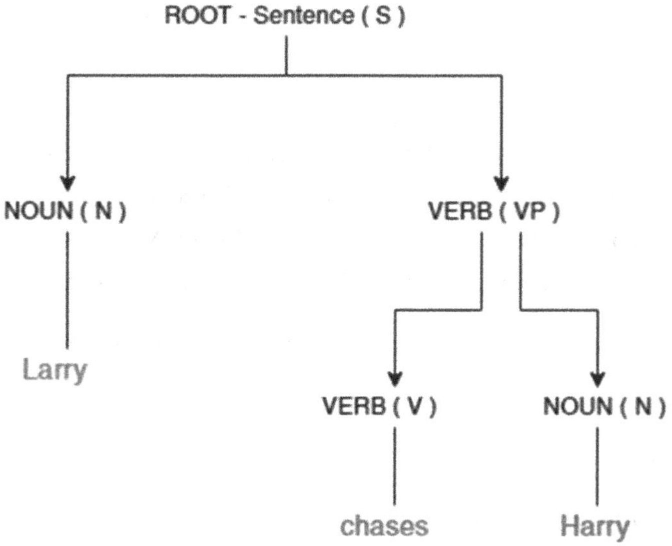

Figure 5-13. *A simple example of constituency parsing*

Dependency Parsing

Dependency parsing is about extracting the syntactic structure of a sentence. It shows the associated set of directed binary grammatical relations that hold among the words in a given sentence. In the spaCy dependency tree, we show a visual representation of the same.

```
print('Dependency Parsing:', nlp.dependency_parse(sentence))
```

```
Dependency Parsing: [('ROOT', 0, 2), ('nsubj', 2, 1), ('case',
5, 3), ('det', 5, 4), ('nmod', 2, 5), ('det', 8, 6), ('nummod',
8, 7), ('nmod:tmod', 2, 8), ('case', 10, 9), ('nmod', 8, 10),
('punct', 2, 11), ('cc:preconj', 13, 12), ('dep', 2, 13),
('dobj', 13, 14), ('case', 16, 15), ('nmod', 13, 16), ('cc',
13, 17), ('conj', 13, 18), ('dobj', 18, 19), ('case', 24, 20),
('det', 24, 21), ('compound', 24, 22), ('compound', 24, 23),
('nmod', 18, 24)]
```

Since Stanford-corenlp is Python in a wrapper on top of the Java-based implementation, you should close the server once the processing is completed. Otherwise, the Java Virtual Memory (JVM) heap will be continuously utilized, hampering other processes in your machine.

```
nlp.close() # Close the server or it will consume much memory.
```

TextBlob

TextBlob is a simple library for beginners in NLP. Although it offers few advanced features like machine translation, it is through a Google API. It is for simply getting to know NLP use cases and on generic datasets. For more sophisticated applications, consider using spaCy or CoreNLP.

POS Tags and Noun Phrase

Similar to the other libraries, TextBlob provides method tags to extract the POS from a given sentence. It also provides the noun_phrase method.

```
#First, the import
from textblob import TextBlob

#create our first TextBlob
s_text = TextBlob("Building Enterprise Chatbot that can
converse like humans")

#Part-of-speech tags can be accessed through the tags property.
s_text.tags

[('Building', 'VBG'),
 ('Enterprise', 'NNP'),
 ('Chatbot', 'NNP'),
 ('that', 'WDT'),
 ('can', 'MD'),
 ('converse', 'VB'),
 ('like', 'IN'),
 ('humans', 'NNS')]

#Similarly, noun phrases are accessed through the noun_phrases
property
s_text.noun_phrases

WordList(['enterprise chatbot'])
```

Spelling Correction

Spelling correction is an exciting feature of TextBlob, which is not provided in the other libraries described in this chapter. The implementation is based on a simple technique provide by Peter Norvig, which is only 70% accurate. The method correct in TextBlob provides this implementation.

111

In the following code, the word "converse" is misspelled as "converce," which the correct method was able to identify correctly. However, it made a mistake in changing the word "Chatbot" to "Whatnot."

```
# Spelling correction
# Use the correct() method to attempt spelling correction.
# Spelling correction is based on Peter Norvig's "How to Write
a Spelling Corrector" as implemented in the pattern library. It
is about 70% accurate

b = TextBlob("Building Enterprise Chatbot that can converce
like humans. The future for chatbot looks great!")
print(b.correct())

Building Enterprise Whatnot that can converse like humans. The
future for charcot looks excellent!
```

Machine Translation

The following code shows a simple example of text translated from English to French. The method translates calls a Google Translate API, which takes an input "to" where we specify the target language to translate. There is nothing novel in this implementation; it is a simple API call.

```
#Translation and language detection
# Google Translate API powers language translation and detection.

en_blob = TextBlob(u'Building Enterprise Chatbot that can
converse like humans. The future for chatbot looks great!')
en_blob.translate(to='fr')

TextBlob("Construire un chatbot d'entreprise capable
de converser comme un humain. L'avenir de chatbot est
magnifique!")
```

Multilingual Text Processing

In this section, we will explore the various libraries and capabilities in handling languages other than English. We find the library spaCy is one of the best in terms of number of languages it supports, which currently stands at more than 50. We will try to perform language translation, POS tagging, entity extraction, and dependency parsing on text taken from the popular French news website www.lemonde.fr/.

TextBlob for Translation

As shown in the example above, we use TextBlob for machine translation so non-French readers can understand the text we process.

The English translation of the text shows that the news is about a match to be played on Friday between two French tennis players, Paire and Mahut, in Roland-Garros.

```
from textblob import TextBlob
```

```
#A News brief from the French news website: https://www.lemonde.fr/
```

```
fr_blob = TextBlob(u"Des nouveaux matchs de Paire et Mahut au
retour du service à la cuillère, tout ce qu'il ne faut pas rater
à Roland-Garros, sur les courts ou en dehors, ce vendredi.")
fr_blob.translate(to='en')
```

```
TextBlob("New matches of Paire and Mahut after the return
of the service with the spoon, all that one should not miss
Roland-Garros, on the courts or outside, this Friday.")
```

POS and Dependency Relations

We use the model fr_core_news_sm from spaCy in order to extract the POS and dependency relation from the given text. To download the model, type

```
python -m spacy download fr_core_news_sm
```

from a command line.

import spacy

```
#Download: python -m spacy download fr_core_news_sm
nlp = spacy.load('fr_core_news_sm')
french_text = nlp("Des nouveaux matchs de Paire et Mahut
au retour du service à la cuillère, tout ce qu'il ne faut
pas rater à Roland-Garros, sur les courts ou en dehors, ce
vendredi.")
```

```
for token in french_text:
    print(token.text, token.pos_, token.dep_)
```

```
Des DET det
nouveaux ADJ amod
matchs ADJ nsubj
de ADP case
Paire ADJ nmod
et CCONJ cc
Mahut PROPN conj
au CCONJ punct
retour NOUN ROOT
du DET det
service NOUN obj
à ADP case
la DET det
cuillère NOUN obl
, PUNCT punct
tout ADJ advmod
ce PRON fixed
qu' PRON mark
il PRON nsubj
ne ADV advmod
```

```
faut VERB advcl
pas ADV advmod
rater VERB xcomp
à ADP case
Roland PROPN obl
- PUNCT punct
Garros PROPN conj
, PUNCT punct
sur ADP case
les DET det
courts NOUN obl
ou CCONJ cc
en ADP case
dehors ADP conj
, PUNCT punct
ce DET det
vendredi NOUN obl
. PUNCT punct
```

Its performance of French POS and dependency relation is entirely accurate. It can identify almost all the VERBS, NOUNS, ADJ, PROPN, and many other tags. Next, let's see how it performs on the entity recognition task.

Named Entity Recognition

The syntax to retrieve the NER remains the same. We see here that the model identified Paire, Mahur, Roland, and Garros as PER entities. We expect the model to give the entity EVENT, since Rolland-Garros is a tennis tournament, as a sports event. Perhaps you could consider training a custom model to extract this entity.

```
# Find named entities, phrases, and concepts
for entity in french_text.ents:
    print(entity.text, entity.label_)
```

```
Paire PER
Mahut PER
Roland PER
Garros PER
```

Noun Phrases

Noun chunks can be extracted using the noun_chunks method provided in the French model from the spaCy library:

```
for fr_chunk in french_text.noun_chunks:
    print(fr_chunk.text, fr_chunk.root.text, fr_chunk.root.dep_,
            fr_chunk.root.head.text)
```

```
et Mahut Mahut conj matchs
du service service obj retour
il il nsubj faut
```

Natural Language Understanding

In recent times, both industry and academia have shown tremendous interest in natural language understanding. This has resulted in an explosion of literature and tools. Some of the major applications of NLU include

- Question answering

- Natural language search

- Web-scale relation extraction

- Sentiment analysis

- Text summarization

- Legal discovery

The above applications can be majorly grouped into four topics:

- **Relation extraction**: Finding the relationship between instances and database tuples. The outputs are discrete values.

- **Semantic parsing**: Parse sentences to create logical forms of text understanding, which humans are good at performing. Again, the output here is a discrete value.

- **Sentiment analysis**: Analyze sentences to give a score in a continuous range of values. A low value means a slightly negative sentiment, and a high score means a positive sentiment.

- **Vector space model**: Create a representation of words as a vector, which then can help in finding similar words and contextual meaning.

We will explore some of the above applications in this section.

Sentiment Analysis

TextBlob provides an easy-to-use implementation of sentiment analysis. The method `sentiment` takes a sentence as an input and provides polarity and subjectivity as two outputs.

Polarity

A float value within the range [-1.0, 1.0]. This scoring uses a corpus of positive, negative, and neutral words (which is called polarity) and detects the presence of a word in any of the three categories. In a simple example, the presence of a positive word is given a score of 1, -1 for negative, and 0 for

117

neutral. We define polarity of a sentence as the average score, i.e., the sum of the scores of each word divided by the total number of words in the sentence.

If the value is less than 0, the sentiment of the sentence is negative and if it is greater than 0, it is positive; otherwise, it's neutral.

Subjectivity

A float value within the range [0.0, 1.0]. A perfect score of 1 means "very subjective." Unlike polarity, which reveals the sentiment of the sentence, subjectivity does not express any sentiment. The score tends to 1 if the sentence contains some personal views or beliefs. The final score of the entire sentence is calculated by assigning each word on subjectivity score and applying the averaging; the same way as polarity.

The TextBlob library internally calls the pattern library to calculate the polarity and subjectivity of a sentence. The pattern library uses SentiWordNet, which is a lexical resource for opinion mining, with polarity and subjectivity scores for all WordNet synsets. Here is the link to the SentiWordNet: https://github.com/aesuli/sentiwordnet.

In the following example, the polarity of the sentence is 0.5, which means it's "more positive" and the subjectivity of 0.4375 means it is "very subjective."

```
s_text = TextBlob("Building Enterprise Chatbot that can
converse like humans. The future for chatbot looks great!")
s_text.sentiment
```

```
Sentiment(polarity=0.5, subjectivity=0.4375)
```

Language Models

The first task of any NLP modeling is to break a given piece of text into tokens (or words), the fundamental unit of a sentence in any language. Once we have the words, we want to find the best numeric representation of the words because machines do not understand words; they need numeric values to perform computation. We will discuss two: Word2Vec (Word to a Vector) and GloVe (Global Vectors for Word Representation). For Word2Vec, a detailed explanation is provided in the next section.

Word2Vec

Word2Vec is a machine learning model (trained with a vast vocabulary of words using the neural network) that produces word embeddings, which are vector representations of words in the vocabulary. Word2vec models are trained to construct the linguistic context of words. We will see some examples in Python using the gensim library to understand what linguistic context means. Figure 5-14 shows the neural network architecture for training the Word2Vec model.

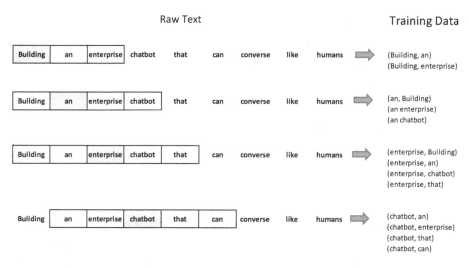

Figure 5-14. *Generating training sample for the neural network using a window size of 2*

A skip-gram neural network model for Word2Vec computes the probability for every word in the vocabulary of being the "nearby word" that we select. Proximity or nearness of words can be defined by a

119

parameter called window size. Figure 5-14 shows the possible pair of words for training a neural network with window size of 2.

Any one of the tools can be used to generate such n-grams. In the following code, we use the TextBlob library in Python to generate the n-grams with a window size of 2.

```
#n-grams
#The TextBlob.ngrams() method returns a list of tuples of n
successive words.
#First, the import
from textblob import TextBlob

blob = TextBlob("Building an enterprise chatbot that can
converse like humans")
blob.ngrams(n=2)

[WordList(['Building', 'an']),
 WordList(['an', 'enterprise']),
 WordList(['enterprise', 'chatbot']),
 WordList(['chatbot', 'that']),
 WordList(['that', 'can']),
 WordList(['can', 'converse']),
 WordList(['converse', 'like']),
 WordList(['like', 'humans'])]
```

In the input sentence, "Building an enterprise chatbot that can converse like humans" is broken into words and with a window size of 2, we take two words each from left and right of the input word. So, if the input word is "chatbot," the output probability of the word "enterprise" will be high because of its proximity to the word "chatbot" in the window size of 2. This is only one example sentence. In a given vocabulary, we will have thousands of such sentences; the neural network will learn statistics from the number of times each pairing shows up. So, if we feed many more training samples like the one shown in Figure 5-14, it will figure out how likely the words "chatbot" and "enterprise" are going to appear together.

Neural Network Architecture

The input vector to the neural network is a one-hot vector representing the input word "chatbot," by storing **1** in the ith position of the vector and **0** in all other positions, where $0 \leq i \leq n$ and n is the size of the vocabulary (set of all the unique words)

In the hidden layer, each word vector of size **n** is multiplied with a feature vector of size, let's say **1000.** When the training starts, the feature vector of size **1000** are all assigned a random value. The result of the multiplication will select the corresponding row in the **n x 1000 matrix** where the one-hot vector has a value of 1.

Finally, in the output layer, an activation function like softmax is applied to shrink the output value to be between 0 and 1. The following equation represents the softmax function, where K is the size of the input vector:

$$\sigma(z)_i = \frac{e^{z_i}}{\sum_{j=1}^{K} e^{z_j}},$$

$$for\ i = 1,\ldots,K\ and\ z = (z_1,\ldots,z_k) \in \mathbb{R}^K$$

So, if the input vector representing "chatbot" is multiplied with the output vector represented by "enterprise," the softmax function will be close to 1 because in our vocabulary, both the words appeared together very frequently.

Neural networks train the network and update the edge weights over many iterations of training. The final set of weights represents learnings. Figure 5-15 shows the neural network architecture to train a Word2Vec model.

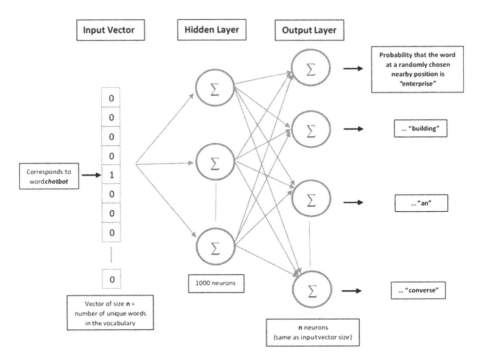

Figure 5-15. *Neural network architecture to train a Word2Vec model*

Using the Word2Vec Pretrained Model

In the following code, we use the pretrained Word2Vec model from a favorite Python library called gensim. Word2Vec models provide a vector representation of words that make various natural language tasks possible, such as identifying similar words, finding synonyms, word arithmetic, and many more. The most popular Word2Vec models are GloVe, CBOW, and skip-gram. In this section, we will use all three models to perform various tasks of NLU.

In the demo, we use the model to perform many syntactic/semantic NLU word tasks.

Step 1: Load the required libraries:

```
from gensim.test.utils import get_tmpfile
from gensim.models import Word2Vec
import gensim.models
```

Step 2: Pick some words from the Amazon Food Review and make a list:

```
review_texts = [['chips', 'WalMart', 'fajitas'],
 ['ingredients', 'tasted', 'crisps', 'Chilling', 'fridge', 'nachos'],
 ['tastebuds', 'tortilla', 'Mexican', 'baking'],
 ['toppings', 'goodness', 'product, 'fantastic']]
```

Step 3: Train the Word2Vec model and save the model to a temporary path. The function Word2Vec trains the neural network on the input vocabulary supplied to it. The following are what the arguments mean:

- review_texts: Input vocabulary to the neural network (NN).

- size: The size of NN layer corresponding to the degree of freedom the algorithm has. Usually, a bigger network is more accurate, provided there is a sizeable dataset to train on. The suggested range is somewhere between ten to thousands.

- min_count: This argument helps in pruning the less essential words from the vocabulary, such as words that appeared once or twice in the corpus of millions of words.

- workers: The function Word2Vec offers for training parallelization, which speeds up the training process considerably. As per the official docs on gensim, you need to install Cython in order to run in parallelization mode.

```
path = get_tmpfile("word2vec.model")

model = Word2Vec(review_texts, size=100, window=5, min_count=1,
workers=4)
model.save("word2vec.model")
```

Note   After installing Cython, you can run the following code to
check if you have the FAST_VERSION of word2vec installed.

---

from gensim.models.word2vec import FAST_VERSION
FAST_VERSION

Step 4: Load the model and get the output word vector using the
attribute wv from the word vector model. The word "tortilla" was one of
the words in the vocabulary. You can check the length of the vector which,
based on the parameter size set during training, is 100; the type of vector is
a numpy array.

```
model = Word2Vec.load("word2vec.model")

vector = model.wv['tortilla']
vector
Out[6]:
array([ 3.4357556e-03,  3.0461869e-03, -1.4244991e-03, -4.6549197e-03,
       -1.8324883e-03,  1.9077188e-04, -1.7216217e-03, -4.5330520e-03,
        3.5653920e-03,  1.4612208e-03,  2.3089715e-03, -2.7617093e-03,
        6.8887050e-04, -5.6756934e-04,  1.1901358e-03,  8.0038357e-04,
        3.0033696e-03, -6.6507862e-05, -4.9998574e-03, -3.6887771e-03,
        2.9287676e-03,  3.6550803e-06, -6.3992629e-04,  4.0531787e-04,
        7.9464359e-04,  3.8370243e-03,  1.5980378e-03,  3.2125930e-03,
       -4.0334738e-03,  2.2513322e-03,  1.6611509e-03, -1.8190868e-03,
        6.9712318e-04,  4.2551439e-03,  1.5517352e-03, -2.8593470e-03,
        3.2627038e-03, -3.9196378e-03,  2.0745063e-04, -2.4973995e-03,
```

```
     -1.9995500e-03,   4.3865214e-03,   2.7636185e-03,   4.1850037e-03,
     -4.4220770e-03,  -1.9331808e-03,  -2.4466095e-03,   3.4395256e-03,
      2.7979608e-03,   7.6796720e-04,  -2.2225662e-03,  -2.3218829e-03,
      1.4716865e-03,   2.5831673e-03,  -2.7626422e-03,  -3.8978728e-03,
     -7.1556045e-05,  -5.0603821e-06,   3.7337472e-03,   1.7892369e-03,
      9.4844203e-04,   4.2107059e-03,   2.0532655e-03,   4.8830300e-03,
      3.9778049e-03,   7.7870529e-04,  -3.0672669e-03,   2.4687734e-03,
     -5.6728686e-04,  -3.1949261e-03,  -3.5277463e-03,  -2.8095061e-03,
      1.9295703e-03,  -2.7000166e-03,   3.8331877e-03,  -3.7821392e-03,
     -2.8160575e-03,  -2.1306602e-03,  -3.4921553e-03,   1.4594033e-03,
      2.9177510e-03,  -7.1679556e-04,  -4.6973061e-03,  -5.6215626e-04,
     -4.7952992e-05,   1.4449660e-03,   3.9334581e-03,  -4.7264448e-03,
      1.3655510e-03,   3.0361500e-03,  -3.9414247e-03,  -2.2786416e-03,
     -2.0382130e-03,   1.2625570e-03,   3.3640184e-03,   3.2833132e-03,
     -4.9897577e-03,   1.3328259e-03,  -3.8654597e-03,  -3.4675971e-03],
    dtype=float32)
```

```
type(vector)
```

```
numpy.ndarray
```

```
len(vector)
```

```
100
```

Step 5: The Word2Vec model we saved in step 3 can be loaded again and we can continue the training on more words using the `train` function in the Word2Vec model.

```
more_review_texts = [['deceptive', 'packaging', 'wrappers'],
 ['texture', 'crispy', 'thick', 'cruncy', 'fantastic', 'rice']]
```

```
model = Word2Vec.load("word2vec.model")
model.train(more_review_texts, total_examples=2,epochs=10)
```

```
(2, 90)
```

Performing Out-of-the-Box Tasks Using a Pretrained Model

One of the useful features of gensim is that it offers several pretrained word vectors from gensim-data. Apart from Word2Vec, it also provides GloVe, another robust unsupervised learning algorithm for finding word vectors. The following code downloads a glove-wiki-gigaword-100 word vector from gensim-data and performs some out-of-the-box tasks.

Step 1: Download one of the pretrained GloVe word vectors using the gensim.downloder module:

```
import gensim.downloader as api
word_vectors = api.load("glove-wiki-gigaword-100")
```

Step 2: Compute the nearest neighbors. As you have seen, word vectors contain an array of numbers representing a word. Now it becomes possible to perform mathematical computations on the vectors. For example, we can compute Euclidean or cosine similarities between any two-word vectors. There are some interesting results that we obtain as a result. The following code shows some of the outcomes.

Figure 5-14 shows an example of how the input data for training the neural network was created by shifting the window of a size 2. In the following example, you will see that "apple" on the Internet is no longer fruit; it has become synonymous with the Apple Corporation and shows many companies like it when we compute a word similar to "apple." The reason for such similarity is because of the vocabulary used for training, which in this case is a Wikipedia dump of close to 6 billion uncased tokens. More such pretrained models are available at `https://github.com/RaRe-Technologies/gensim-data`.

In the second example, when we find similar words to "orange," we obtain words corresponding to colors like red, blue, purple, and fruits like lemon, which is a citrus fruit like an orange. Such relations are easy for humans to understand. However, it is exciting how the Word2Vec model can crack it.

```
result = word_vectors.most_similar('apple')
print(result)
```

```
[('microsoft', 0.7449405789375305), ('ibm',
0.6821643114089966), ('intel', 0.6778088212013245),
('software', 0.6775422096252441), ('dell', 0.6741442680358887),
('pc', 0.6678153276443481), ('macintosh', 0.66175377368927),
('iphone', 0.6595611572265625), ('ipod', 0.6534676551818848),
('hewlett', 0.6516579389572144)]
```

```
result = word_vectors.most_similar('orange')
print(result)
```

```
[('yellow', 0.7358633279800415), ('red', 0.7140780687332153),
('blue', 0.7118035554885864), ('green', 0.7111418843269348),
('pink', 0.6775072813034058), ('purple', 0.6774232387542725),
('black', 0.6709616184234619), ('colored', 0.665260910987854),
('lemon', 0.6251963376998901), ('peach', 0.616862416267395)]
```

Step 3: Identify linear substructures. The relatedness of two words is easy to compute using the similarity or distance measure, whereas to capture the nuances in a word pair or sentences in a more qualitative way, we need operations. Let's see the methods that the gensim package offers to accomplish this task.

Word Pair Similarity

In the following example, we find a similarity between a word pair. For example, the word pair ['sushi', 'shop'] is more similar to the word pair ['japanese', 'restaurant'] than to ['Indian', 'restaurant'].

```
sim = word_vectors.n_similarity(['sushi', 'shop'], ['indian',
'restaurant'])
print("{:.4f}".format(sim))
```

0.6438

```
sim = word_vectors.n_similarity(['sushi', 'shop'], ['japanese',
'restaurant'])
print("{:.4f}".format(sim))
```

0.7067

Sentence Similarity

We can also find distance or similarity between two sentences. gensim offers a distance measure called Word Mover's distance, which has proved quite a useful tool in finding out the similarity between two documents that contain many sentences. The lower the distance, the more similar the two documents. Word Mover's distance underneath uses the word embeddings generated by the Word2Vec model to first understand the concept of the query sentence (or document) and then find all the similar sentences or documents. For example, when we compute the Mover's distance between two unrelated sentences, the distance is high compared to when we compare two sentences that are contextually related.

In the first example, sentence_one talks about diversity in Indian culinary art, and sentence_two specifically talks about the food in Delhi. In the second example, sentence_one and sentence_two are unrelated, so we get a higher Movers distance than the first example.

```
sentence_one = 'India is a diverse country with many culinary
art'.lower().split()
sentence_two = 'Delhi offers many authentic food'.lower().split()

similarity = word_vectors.wmdistance(sentence_one, sentence_two)
print("{:.4f}".format(similarity))
```

4.8563

```
sentence_one = 'India is a diverse country with many culinary
art'.lower().split()
sentence_two = 'The all-new Apple TV app, which brings together
all the ways to watch TV into one app'.lower().split()

similarity = word_vectors.wmdistance(sentence_one, sentence_two)
print("{:.4f}".format(similarity))
```

5.2379

Arithmetic Operations

Even more impressive is the ability to perform arithmetic operations like addition and subtraction on the word vector to obtain some form of linear substructure because of the operation. In the first example, we compute *woman + king – man,* and the most similar word to this operation is *queen.* The underlying concept is that man and woman are genders, which may be equivalently specified by other words like queen and king. Hence, when we take out the man from the addition of woman and king, the word we obtain is queen. GloVE word representation provides few examples here: `https://nlp.stanford.edu/projects/glove/`.

Similarly, the model is good at picking up concepts like language and country. For example, when we add French and Italian, it gives Spanish, which is a language spoken in a nearby country, Spain.

```
result = word_vectors.most_similar(positive=['woman', 'king'],
negative=['man'])
print("{}: {:.4f}".format(*result[0]))
```

```
queen: 0.7699
```

```
result = word_vectors.most_similar(positive=['french',
'italian'])
print("{}: {:.4f}".format(*result[0]))
```

```
spanish: 0.8312
```

```
result = word_vectors.most_similar(positive=['france', 'italy'])
print("{}: {:.4f}".format(*result[0]))
```

```
spain: 0.8260
```

Odd Word Out

The model adapts to find words that are out of context in a given sequence of words. The way it works is the method doesnt_match computes the center point by taking the mean of all the word vectors in a given list of words and finding the cosine distance from the center point. The word with the highest cosine distance is returned as an odd word that does not fit in the list.

In the following two examples, the model was able to pick the food pizza as an odd word out from the list of countries. Similarly, in the second example, the model picked up the Indian Prime Minister Modi from the list of all US Presidents.

```
print(word_vectors.doesnt_match("india spain italy pizza".split()))
```

```
pizza
```

```
print(word_vectors.doesnt_match("obama trump bush modi".split()))
```

```
modi
```

Language models like Word2Vec and GloVe are compelling in generating meaningful relationships between words, which comes naturally to a human because of our understanding of languages. It is an excellent accomplishment for machines to be able to perform at this level of intelligence in understanding the use of words in various syntactic and semantic forms.

fastText Word Representation Model

Similar to gensim, fastText also provides many pretrained word embedding models. Its fast and efficient processing makes it a very popular library for text classification and tasks related to word representation such as finding similar text. The models in fastText use subword information, all the substrings contained in a word between the minimum size (minn) and the maximal size (maxn), which give better performance.

import fasttext

```
# Skipgram model
model_sgram = fasttext.train_unsupervised('dataset/amzn_food_
review_small.txt', model='skipgram')

# or, cbow model
model_cbow = fasttext.train_unsupervised('dataset/amzn_food_
review_small.txt', model='cbow')

print(model_sgram['cakes'])

[ 0.00272718  0.01386657  0.00484232 -0.01444803  0.00204112
  0.00787148
 -0.00759551  0.00263086 -0.01182229 -0.00530771 -0.02338764
  0.01398039
  0.00218989  0.0154795  -0.01450872 -0.01040525 -0.00762093
 -0.01090531
```

```
    0.00802671 -0.02447837   0.00507444   0.01049152 -0.00054866
   0.01148072
  -0.02119654 -0.01219683   0.00658704 -0.00171852   0.01495257
   0.00328717
  -0.01289422   0.01350378 -0.01774059   0.01281367   0.00123221
 -0.01672287
  -0.00940464 -0.01039432 -0.00618952   0.01418524 -0.03802125
   0.00976629
    0.01477897   0.01039862   0.02141832 -0.01620196   0.00617392
 -0.01073407
  -0.00289557 -0.00856876 -0.00785293 -0.01535104   0.00439641
 -0.00760364
    0.00825184   0.03034449 -0.00980587   0.01319963 -0.00710381
   0.00040615
  -0.0074836    0.01588171   0.03172321   0.00821354   0.00569351
 -0.00976394
  -0.00666583   0.00810414 -0.00969361 -0.00378272   0.00782087
   0.01669582
    0.01114488   0.00669733 -0.0053518   -0.0059374   -0.00554186
   0.01869696
    0.01529924 -0.00877811   0.03367095   0.01772366   0.0037948
   0.01354953
  -0.0086841    0.01565165 -0.0031147    0.00028975 -0.00047118
 -0.00779429
  -0.00646258   0.00798804   0.04278774 -0.00381226 -0.01868668
 -0.01809955
  -0.02041707 -0.00328311 -0.01909724 -0.01288191]

print(model_sgram.words)

['the', 'I', 'a', 'and', 'to', '</s>', 'of', 'for', 'it', 'in',
'is', 'was', 'are', 'not', 'this', 'that', 'but', 'on', 'my',
```

'have', 'as', 'they', 'like', 'you', 'great', 'This', 'so',
'them', 'than', 'body', 'soap', 'just', 'The', 'very', 'find',
'with', 'taste', 'cake', 'what', 'these', 'had', 'when', 'buy',
'get', 'be', 'It', 'sprinkles', 'from', 'really', "it's",
'Great', 'other', 'Giovanni', 'best', 'we', 'good', 'all',
'were', 'out', 'wash', 'one', 'only', 'their', 'make', 'about',
'or', 'color', 'bag', '/><br', 'some', 'These', 'using',
'bought', 'tried', 'your', 'more', 'same', 'any', "I've",
'also', 'love', 'has', 'washes']

Similar to the examples discussed in this section, using either the skip-gram or CBOW model, various tasks can be performed. We can evaluate the performance to choose the best model for our final implementation.

It's possible to use the fastText model from within the gensim library by importing the fastText module:

```
from gensim.models.fasttext import FastText
```

Information Extraction Using OpenIE

The Open Information Extractor (OpenIE) annotator extracts open-domain relation triples representing subject, predicate, and object, often called a triplet. OpenIE can be a useful tool when there is minimal training data available.

There is no stable implementation of OpenIE in Python. In order to use OpenIE provided by CoreNLP library, download corenlp and from the command line, type cd into the CoreNLP directory. Then run the following command. Note that this process requires the right amount of RAM. In the following code, we set 2GB RAM for running this process. Otherwise, the JVM might throw an out of memory error.

```
java -mx2g -cp "*" edu.stanford.nlp.naturalli.OpenIE
```

Once the above command runs, it takes one sentence as input. Provide a sentence of your choice. To reproduce the same result as shown in Table 5-3, use the following example sentence:

```
Narendra Modi is an Indian politician serving as the 14th and
current Prime Minister of India since 2014
```

Table 5-4 shows the possible triplets from the given sentence. At first, many triplets may look the same. On careful examination, you can see that all the objects are all unique using the subject "Narendra Modi" or "Modi" and predicate or the relation "is."

Table 5-4. *The Possible Triplets from the Example Sentence Using OpenIE*

S.No	Subject	Predicate	Object
1	Narendra Modi	is	politician serving as 14th Prime Minister
2	Narendra Modi	is	Indian politician serving as 14th Prime Minister
3	Narendra Modi	is	politician serving as Prime Minister
4	Narendra Modi	is	Politician
5	Modi	is	Indian
6	Narendra Modi	is	Indian politician serving as 14th Prime Minister of India
7	Narendra Modi	is	Indian politician serving as Prime Minister
8	Narendra Modi	is	Indian politician serving as Prime Minister of India since 2014

(continued)

Table 5-4. (*continued*)

S.No	Subject	Predicate	Object
9	Narendra Modi	is	Indian politician serving as Prime Minister since 2014
10	Narendra Modi	is	politician serving as Prime Minister of India since 2014
11	Narendra Modi	is	politician serving as 14th Prime Minister of India since 2014
12	Narendra Modi	is	politician serving as 14th Prime Minister since 2014
13	Narendra Modi	is	Indian politician serving as 14th Prime Minister since 2014
14	Narendra Modi	is	politician serving as Prime Minister of India
15	Narendra Modi	is	politician serving as Prime Minister since 2014
16	Narendra Modi	is	Indian politician
17	Narendra Modi	is	Indian politician serving as Prime Minister of India
18	Narendra Modi	is	Indian politician serving since 2014
19	Narendra Modi	is	politician serving as 14th Prime Minister of India
20	Narendra Modi	is	politician serving since 2014
21	Narendra Modi	is	Indian politician serving as 14th Prime Minister of India since 2014

Topic Modeling Using Latent Dirichlet Allocation

Topic modeling is one of the typical applications of understanding natural language. Given a collection of documents, we can draw an "abstract topic" that represents all the docs in the collection. Latent Dirichlet allocation (LDA) is a favorite statistical model used for topic modeling. It helps in discovering the semantic structures in a given text.

In this section, for a demonstration, we will use three example reviews from the Amazon Fine Food review dataset to train an LDA model. We will see one other example of topic modeling using additional tools like spaCy, NLTK, and gensim in the "Applications" sections.

Collection of Documents

Three reviews from the dataset are assigned to a variable named documents. We expect the topics to have words like "chips," "fajitas," and "crisps" as these three reviews seem to be talking about "corn chips." We are not much concerned about the sentiment in the review.

```
documents = ["I consume about a jar every two weeks of this,
either adding it to fajitas or using it as a corn chip dip,"
            "As soon as I tasted one and it tasted like a corn
            chip I checked the ingredients",
            "I found these crisps at our local WalMart &
            figured I would give them a try"
]
```

Loading Libraries and Defining Stopwords

As a first preprocessing step, we remove all the stopwords from the given text. For a simple implementation, we have only defined a few stopwords in a list:

```python
# Import pretty printer
from pprint import pprint
from collections import defaultdict
stoplist = set('for a of the and to in'.split())
```

Removing Common Words and Tokenizing

Using the stopwords in the list above, we run through a for-loop to remove the words. Note that this is a simple implementation and is not the most efficient way of removing the stopwords.

```python
# Remove common words and tokenize
texts = [
    [word for word in document.lower().split() if word not in
    stoplist]
    for document in documents
 ]
```

Removing Words That Appear Infrequently

Now that we have removed the stopwords, we compute the frequency of occurrence of each word in the document collection. Again, we implement this using a simple two for-loop structure that reads each word in the document and increments the count whenever we encounter a word more than once.

```python
# Remove words that appear only once
frequency = defaultdict(int)
```

```
for text in texts:
    for the token in text:
        frequency[token] += 1

texts = [
    [token for token in text if frequency[token] > 1]
    for text in texts
]

pprint(texts)

[['i', 'it', 'it', 'as', 'corn', 'chip'],
 ['as', 'as', 'i', 'tasted', 'it', 'tasted', 'corn', 'chip', 'i'],
 ['i', 'i']]
```

Now we see the words that occur more than once. For our example, it seems like there are not many words with more than one occurrence. We expect the model not to perform very well. However, let's still go ahead with training the model.

Saving the Training Data as a Dictionary

The gensim library provides the method Dictionary, which stores the tokens into a dictionary. We save the tokens extracted from the review in the review.dict file on disk.

```
from gensim import corpora
dictionary = corpora.Dictionary(texts)
dictionary.save('review.dict')
print(dictionary)

Dictionary(6 unique tokens: ['as', 'chip', 'corn', 'i', 'it']...)
```

```
print(dictionary.token2id)
```

```
{'as': 0, 'chip': 1, 'corn': 2, 'i': 3, 'it': 4, 'tasted': 5}
```

```
new_doc = "tasty corn"
new_vec = dictionary.doc2bow(new_doc.lower().split())
print(new_vec)
```

```
[(2, 1)]
```

Generating the Bag of Words

Words in the dictionary can be converted to a bag-of-words (BOW) representation using the method doc2bow. The BOW can then be serialized using MmCorpus and stored as review.mm. Another popular approach is to represent the words in an n-gram, where the text is ordered rather than unordered in case BOW. N-grams helps to find the cooccurrence among words.

```
corpus = [dictionary.doc2bow(text) for text in texts]
corpora.MmCorpus.serialize('review.mm', corpus)
```

Training the Model Using LDA

Finally, using the bag-of-words dictionary of words, we train the latent Dirichlet allocation model. LDA is a generative statistical model where, given an input variable X and target variable Y, the model based on joint probability is $X * Y$, $P(X, Y)$. LDA is a favorite machine learning model widely used in topic modeling. Each document (in our example, each review) is a mixture of various topics, where each document is assigned a set of topics by LDA.

For example, the LDA model may assign a topic for a review (documents), something like "corn chips" related. This topic has probabilities of generating various words like "crispy," "tasty," and so on.

```python
from gensim import models
tfidf = models.TfidfModel(corpus)
corpus_tfidf = tfidf[corpus]
lsi = models.LsiModel(corpus_tfidf, id2word=dictionary, num_
topics=2)
corpus_lsi = lsi[corpus_tfidf]
lsi.print_topics(2)
```

```
[(0,
  '0.556*"it" + 0.542*"tasted" + 0.428*"as" + 0.328*"chip" +
  0.328*"corn" + 0.000*"i"'),
 (1,
  '-0.804*"tasted" + 0.528*"it" + 0.190*"corn" + 0.190*"chip" +
  0.041*"as" + 0.000*"i"')]
```

The gensim library provides a method called LsiModel(), which trains an LDA model. Latent semantic indexing (LSI) is used in the context of LDA's application in information retrieval.

In the model above, we set num_topics = 2, asking the model to generate two topics. The following two topics give weight to "corn" and "chip," which seems to be the topic from the three reviews we used for training.

1. **Topic 1:** 0.556*"it" + 0.542*"tasted" + 0.428*"as" + ***0.328*"chip" + 0.328*"corn"*** + 0.000*"i"

2. **Topic 2:** -0.804*"tasted" + 0.528*"it" + ***0.190*"corn"*** + ***0.190*"chip"*** + 0.041*"as" + 0.000*"i"

Note that a more accurate model would need plenty of data for training and perhaps many more interesting topics might evolve.

Natural Language Generation

Natural language generation is a subfield of NLP and computational linguistics that can produce understandable human text in various languages. The ability to use the language representation and knowledge of the domain to produce documents, explanations, help messages, reports, and even poems makes NLG the most researched area just now.[1] In the future, NLG will play a vital role in human-computer interfaces.

The significant differences between NLU and NLG are that NLP maps sentences into internal semantic representations (called parsing in NLU systems), whereas NLG maps the semantic representation into surface sentences (called realization in NLG systems). Both of these types of mapping are achieved through bidirectional grammar, which uses a declarative representation of a language's grammar.

We will demonstrate NLG applications using Python- and Java-based libraries like markovify and simpleNLG. We will also use a deep learning model for text generation. Such deep learning models are behind the popular use cases where machines are writing poems or generating musical notes given a sizeable corpus of data.

Some popular applications of NLG are

- Automating the documentation of code and procedures

- Generating reports from financial data or annual reports

- Summarizing graphical reports and numbers from tabular data

- Generating discharge summaries and pathology reports

- Helping meteorologists compose weather forecasts

---

[1]https://web.stanford.edu/class/cs224n/slides/cs224n-2019-lecture15-nlg.pdf

There are many more use cases that are evolving quickly, especially with the emerging sophistication of deep learning algorithms and increasing computation power of machines.

Markov Chain-Based Headline Generator

The Markov chain model is a stochastic model describing the sequence of possible events in which the probability of each event depends only on the state achieved in the previous state. Markov chains statistically model random processes. Markov chains are defined by transition probabilities and states, where a process moves from one state to another based on a preset probability value.

Markov chain models are mathematically robust methods that give superior results if modeled correctly. Unlike many machine learning algorithms, which work in a brute-force approach, Markov chains need a diligent design to model a stochastic process.

The following are some applications of Markov chains:

- Computer simulation of numerous real-world phenomena such as weather modeling, stock market fluctuations, and water flow in a dam

- Biological modeling like population processes

- Algorithmic music composition

- Model boards game like Snakes and Ladders or Hi Ho! Cherry-O

- Population genetics to describe changes in gene frequencies in small populations affected by genetic drift

Let's use the markovify library from Python to generate some headlines.

Loading the Library

Load the libraries such as pandas and markovify. We use pandas to read and process the CSV files from an ABC news dataset. The markovify library, a simple Markov chain generator, generates random text.

```
#Loading required packages
import pandas as pd # data processing, CSV file I/O (e.g.
pd.read_csv)
import markovify #Markov chain generator
```

Loading the File and Printing the Headlines

Read the ABC news dataset from a CSV file using the read_csv() method and print the top three news headlines. The dataset contains over 15 years of news headlines published by the Australian Broadcasting Corp. The dataset contains more than 1 million news headlines. The dataset is available for download from www.kaggle.com/therohk/million-headlines/data. See Figure 5-16.

```
#Reading input text file

Input_text = pd.read_csv('data/abcnews-date-text.csv')
Input_text.head(3)
```

	publish_date	headline_text
0	20030219	aba decides against community broadcasting lic...
1	20030219	act fire witnesses must be aware of defamation
2	20030219	a g calls for infrastructure protection summit

Figure 5-16. *Output*

Building a Text Model Using Markovify

Markovify offers a method called NewlineText to take the input as headline_text from the dataset and a paramet value of state_size as 2. This method works best with large and well-punctuated text. Each word is a state in a sequence, and the probability measures which word is likely to come next after the occurrence of one word.

#Building the text model with markovify

```
text_model = markovify.NewlineText(input_text.headline_text,
state_size = 2)
```

Generating Random Headlines

Once you markovify the text, the model can be used to make sentences using the method make_sentence(). This method randomly generates sentences using the model build using the Newlinetext() method in the markovify library. Many of the examples in the random sentences below form meaningful headlines.

```
#Generate random text
# Print ten randomly-generated sentences using the built model
for i in range(10):
    print(text_model.make_sentence())
```

```
coalitions grand plan for fertiliser price hurting jewellers
police seek 18 over brawl outside black magic rape sentencing
dojokvic eases past querrey; murray wins at ascot
life at the waca
police shoot terrorism suspect to undergo mental check
beazley stands by online petition to stop roxon
ocean queen docks in fremantle with yacht damaged in downpour
ministerial clout needed to beat deadline
port macquarie waterfront land row
```

SimpleNLG

Unlike Makov chains, which generate random text based on state transition probabilities, SimpleNLG offers a utility to generate sentences in English that are grammatically correct. It's written in Java for NLG. To generate the sentences, we specify the content of the sentence and encode this information in SimpleNLG syntax, which in turn generates the grammatically correct sentences based on the grammar specifications. Significant tasks that SimpleNLG perform are

- **Orthography**: This refers to the conventions for writing languages. It includes capitalization, whitespaces in sentences, and paragraphs, punctuation, emphasis, and hyphenation.

- **Morphology**: The study of words, their formation, and relationship with other words in the same language. It analyzes the structure of words and parts of words, such as stems, root words, prefixes, and suffixes.

- **Simple grammar**: Ensures grammatical correctness like noun-verb agreement and creating well-formed verb groups (e.g. "does not play").

In the terminology of NLG, SimpleNLG is a realizer for simple grammar. It can be useful for creating documentations and reports that need to use grammatically correct sentences. The demonstration in this section uses nglib, which is a Python library that is mainly a wrapper around SimpleNLG.

Loading the Library

Load the SimpleNLG realizer from the nglib library. Set the host parameter in the `Realiser()` class as `nlg.kutlak.info`. Next, we define methods for the various tasks SimpleNLG is capable of doing.

```
import logging
from nlglib.realisation.simplenlg.realisation import Realiser
from nlglib.microplanning import *
realise = Realiser(host='nlg.kutlak.info')
```

Tense

The method named `tense()` defines a clause and the tense we would like to convert it to. In the following code, the clause declares "Subject," "Predicate" (or relationship), and "Object," and then we set the attribute TENSE in the clause object to PAST and FUTURE separately.

```
def tense():
    c = Clause('Harry', 'bought', 'these off amazon')
    c['TENSE'] = 'PAST'
    print(realise(c))
    c['TENSE'] = 'FUTURE'
    print(realise(c))
```

Harry bought these off amazon.
Harry will buy these off amazon.

Negation

Similar to the method tense(), we define the method negation(), which again takes a triplet and creates a negation of the sentence.

```
def negation():
    c = Clause('Harry', 'bought', 'these off amazon')
    c['NEGATED'] = 'true'
    print(realise(c))
```

Harry does not buy these off amazon.

Interrogative

We can also generate sentences with YES or NO kinds of interrogative sentences or questions like WHO. The following code shows two examples. Note that WHO doesn't go well with "Harry" in the example.

```
def interrogative():
    c = Clause('Harry', 'bought', 'these off amazon')
    c['INTERROGATIVE_TYPE'] = 'YES_NO'
    print(realise(c))
    c['INTERROGATIVE_TYPE'] = 'WHO_OBJECT'
    print(realise(c))
```

147

```
Does Harry buy these off amazon?
Who does Harry buy?
```

Complements

In a given clause, certain complement phrases can also be added. In the following code, we show two complement phrases added to the main clause. The good part of SimpleNLP is that it can form grammatically correct sentences given the clause and complements.

```python
def complements():
    c = Clause('Harry', 'bought', 'these off amazon',
                complements=['on first day of sales',
                'despite high price'])
    print(realise(c))
```

```
Harry buys these off amazon on first day of sales despite high
price.
```

Modifiers

In the following code, we first add the adjective to the subject or the noun and then we add an adverb to the verb. In the example, the adjective "impulsive" is added to the noun "Harry" and the adverb "quickly" is added to the verb "buys." Both the adjective and adverb are called modifiers. Observe that the grammar of the sentence is still intact.

```python
def modifiers():
    subject = NP('Harry')
    verb = VP('bought')
    objekt = NP('these', 'off','amazon')
    subject += Adjective('Impulsive')
    c = Clause()
    c.subject = subject
```

```
c.predicate = verb
c.object = objekt
print(realise(c))
verb += Adverb('quickly')
c = Clause(subject, verb, objekt)
print(realise(c))
```

```
Impulsive Harry buys this off amazon.
Impulsive Harry quickly buys this off amazon.
```

Prepositional Phrases

Prepositional phrases using "at," "on," "in," and "by" are easy to add to the clause using SimpeNLP. In a prepositional phrase, you can also define the noun term separately and it structures it appropriately based on the grammar.

```
def prepositional_phrase():
    c = Clause('Harry', 'bought', 'these off amazon')
    c.complements += PP('by', 'surprise')
    print(realise(c))
    c = Clause('Harry', 'bought', 'these off amazon')
    c.complements += PP('for', NP('Eva'))
    print(realise(c))
```

```
Harry buys these off amazon by surprise.
Harry buys these off amazon for Eva.
```

Coordinated Clauses

In a coordinated clause, two or more sentences (clauses) can be combined to make one sentence. In the following code, two clauses are combined using a conjunction. And each clause can have its own structure. For example, in one clause, "He likes jeans," we use the PRESENT tense and in the second clause, "He will return t-shirt," we use the FUTURE tense.

```
def coordinated_clause():
    s1 = Clause('Harry', 'buy', 'these off amazon',
    features={'TENSE': 'PAST'})
    s2 = Clause('he', 'like','jeans', features={'TENSE':
    'PRESENT'})
    s3 = Clause('he', 'return', 't-shirt', features={'TENSE':
    'FUTURE'})
    c = s1 + s2 + s3
    c = CC(s1, s2, s3)
    print(realise(s1))
    print(realise(s2))
    print(realise(s3))
    print(realise(s1 + s2))
    print(realise(c))
```

```
Harry bought these off amazon.
He likes jeans.
He will return t-shirt.
Harry bought these off amazon and he likes jeans
Harry bought these off amazon and he likes jeans and he will
return t-shirt
```

Subordinate Clauses

We can introduce a conjunction in a clause with a COMPLIMENTIZER like "because" and put the sentence in the past tense. We call this a subordinate clause.

```
def subordinate_clause():
    p = Clause('Harry', 'like', 'amazon')
    q = Clause('product', 'is', 'good')
```

```
q['COMPLEMENTISER'] = 'because'
q['TENSE'] = 'PAST'
p.complements += q
print(realise(p))
```

```
Harry likes amazon because product was good.
```

Main Method

The main method calls the methods we created above if we need to run all the code together at once.

```
def main():
    c = Clause('Harry', 'bought', 'these off amazon')
    print(realise(c))
    tense()
    negation()
    interrogative()
    complements()
    modifiers()
    prepositional_phrase()
    coordinated_clause()
    subordinate_clause()
```

Printing the Output

Let's print the output of all the methods together here in the main method:

```
if __name__ == '__main__':
    logging.basicConfig(level=logging.WARNING)
    main()
```

```
Harry buys these off amazon.
Harry bought these off amazon.
```

Harry will buy these off amazon.
Harry does not buy these off amazon.
Does Harry buy these off amazon?
Who does Harry buy?
Harry buys these off amazon on first day of sales despite high price.
Impulsive Harry buys this off amazon.
Impulsive Harry quickly buys this off amazon.
Harry buys these off amazon by surprise.
Harry buys these off amazon for Eva.
Harry bought these off amazon.
He likes jeans.
He will return t-shirt.
Harry bought these off amazon and he likes jeans
Harry bought these off amazon and he likes jeans and he will return t-shirt
Harry likes amazon because product was good.

As you can see, SimpleNLG offers an easy-to-use syntax to generate grammatically correct sentences in English programmatically. Next, let's dive into a deep learning model to generate the next words given a piece of text. Unlike SimpleNLG, we are not sure if we'll get a grammatically correct sentence in such a deep learning model.

Deep Learning Model for Text Generation

Text generation using deep learning is built for language models and applications like speech-to-text, conversational chatbots, and text summarizations. Such language models predict the occurrences of a word based on the previous sequence of words. Many deep learning network architectures such as recurrent neural networks are available for language modeling.

RNNs are deployed in a variety of applications like speech recognition, language modeling, translation, image captioning, and many more. Figure 5-17 shows how the hidden layers in RNNS are stacked up in a sequence of the chain. The rolled and unrolled versions help in understanding how internal processing happens.

In the demonstration code, we use a deep learning model called the long short-term memory model (LSTM). LSTMs are a particular type of RNN, capable of learning long term dependency, which RNNs are not very good at leaning. One significant difference in the capability between RNNs and LSTM is the ability to understand the context of a word, which might not come from its immediate predecessor but instead could come from a couple of words ahead. For example, if we are trying to predict the next word based on the previous ones, in "I grew up in France… I speak fluent French," the context for French is present further back in the sentence than just the previous word. Figure 5-17 shows the RNN network after unrolling. Observe that each of the neural network chunks labelled N are exactly the same.

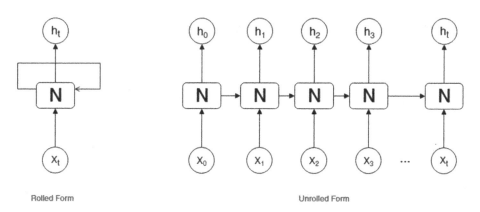

Figure 5-17. *RNN architecture in rolled and unrolled forms*

Although RNNs are capable of picking up such long-term dependency in sentences, they requires a careful selection of parameters, which is often difficult in many practical problems. This is where LSTMs come to the rescue.

Figure 5-18 shows the architecture of the LSTM network.

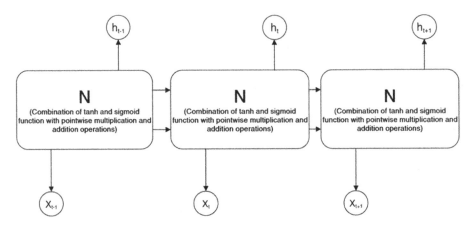

Figure 5-18. *Architecture of an LSTM*

There are four major parts in LSTMs networks:

- **Cell state**: The line that runs through from the top with few direct interactions like pairwise multiplication and addition, which could add or remove any information from the cell state.

- **Forget gate layer**: Gates are a mechanism by which the LSTMs control how much of the information should be passed through the cell state. Here a sigmoid function is used, which has an output value between 0 to 1. If the value is 1, it means let everything pass; 0 means do not let anything pass.

- **Input gate layer:** The sigmoid layer called an input gate layer decides which values we will update.

- **Tanh layer:** The tanh activation function layer creates a vector of new candidate values given the input and hidden state values from the previous time step.

Loading the Library

Load the required libraries from Keras, an open source neural network library built in Python. It's a popular library used for building deep learning models in standalone mode or on top of frameworks like TensorFlow, CNTK, and Theano. It provides fast experimentation with deep learning models with user-friendly, modular, and extensible syntax and structure.

```
from keras.preprocessing.sequence import pad_sequences
from keras.layers import Embedding, LSTM, Dense, Dropout
from keras.preprocessing.text import Tokenizer
from keras.callbacks import EarlyStopping
from keras.models import Sequential
import keras.utils as ku
import numpy as np
```

Defining the Training Data

We took a review from the Amazon Fine Food review dataset. However, more data would get better results.

```
review_data = ""Chilling in the fridge seems to boost the
flavor even more;
and using them, rather than corn chips, to make nachos will
have your tastebuds
singing like Janet Jackson but without any of the associated
wardrobe risks."
```

Data Preparation

Let's define a method called `dataset_preparation` to perform the following major tasks:

1. Convert the input review text into lowercase and split the review into sentences split by a newline character, \n. The split function created three sentences in the corpus. The following is the result of the operation:

   ```
   corpus = review_data.lower().split("\n")
   print(corpus)
   ```

   ```
   ['chilling in the fridge seems to boost the flavor
   even more; ', 'and using them, rather than corn chips,
   to make nachos will have your tastebuds ', 'singing
   like janet jackson but without any of the associated
   wardrobe risks.']
   ```

2. Tokenize the input reviews from the dataset. Use the Keras `fit_on_text()` method. The method internally represents the words in a dictionary with each word getting an index based on the frequency of its occurrence. So, if the word "the" in our review text appears the most, it gets an internal representation with the least index value like `word_index["the"] = 0`. In our review, except for the words "the" and "to," all other words appear just once. The following is the output of the operation:

   ```
   review_tokenizer.fit_on_texts(corpus)
   print(review_tokenizer.word_index)
   ```

   ```
   {'the': 1, 'to': 2, 'chilling': 3, 'in': 4, 'fridge':
   5, 'seems': 6, 'boost': 7, 'flavor': 8, 'even': 9,
   ```

```
'more': 10, 'and': 11, 'using': 12, 'them': 13,
'rather': 14, 'than': 15, 'corn': 16, 'chips': 17,
'make': 18, 'nachos': 19, 'will': 20, 'have': 21,
'your': 22, 'tastebuds': 23, 'singing': 24, 'like': 25,
'janet': 26, 'jackson': 27, 'but': 28, 'without': 29,
'any': 30, 'of': 31, 'associated': 32, 'wardrobe': 33,
'risks': 34}
```

3. Transform each word in the review into a
    sequence of integers. Each word gets the integer
    value corresponding to the index obtained using
    fit_on_text(). The following is the output of

    ```
    token_list = review_tokenizer.texts_to_
    sequences([line])[0]
    print(token_list)
    ```

    ```
    [3, 4, 1, 5, 6, 2, 7, 1, 8, 9, 10]
    [11, 12, 13, 14, 15, 16, 17, 2, 18, 19, 20, 21, 22, 23]
    [24, 25, 26, 27, 28, 29, 30, 31, 1, 32, 33, 34]
    ```

    Note that we generate the index using fit_
    on_text() once and could use the texts_to_
    sequence() as many times we want. The integer
    value assigned to each word makes the computation
    in neural network feasible. This approach is superior
    to assigning a random number to each word at the
    start of the neural network training.

4. Generate n-gram values using the integer sequence
    for each sentence in the corpus. In each iteration of
    the *for* loop, the list input_review_sequences gets
    updated. In the final output, all possible n-grams of
    length 1 to len(token_list) get generated.

```
for line in corpus:
    token_list = review_tokenizer.texts_to_
    sequences([line])[0]
    for i in range(1, len(token_list)):
        n_gram_sequence = token_list[:i+1]
        input_review_sequences.append(n_gram_
        sequence)
    print(input_review_sequences)
```

Iteration 1:
[[3, 4], [3, 4, 1], [3, 4, 1, 5], [3, 4, 1, 5, 6], [3,
4, 1, 5, 6, 2], [3, 4, 1, 5, 6, 2, 7], [3, 4, 1, 5, 6,
2, 7, 1], [3, 4, 1, 5, 6, 2, 7, 1, 8]]

Iteration 2:
[[3, 4], [3, 4, 1], [3, 4, 1, 5], [3, 4, 1, 5, 6], [3,
4, 1, 5, 6, 2], [3, 4, 1, 5, 6, 2, 7], [3, 4, 1, 5,
6, 2, 7, 1], [3, 4, 1, 5, 6, 2, 7, 1, 8], [3, 4, 1, 5,
6, 2, 7, 1, 8, 9], [3, 4, 1, 5, 6, 2, 7, 1, 8, 9, 10],
[11, 12], [11, 12, 13], [11, 12, 13, 14], [11, 12, 13,
14, 15], [11, 12, 13, 14, 15, 16], [11, 12, 13, 14, 15,
16, 17], [11, 12, 13, 14, 15, 16, 17, 2], [11, 12, 13,
14, 15, 16, 17, 2, 18], [11, 12, 13, 14, 15, 16, 17, 2,
18, 19], [11, 12, 13, 14, 15, 16, 17, 2, 18, 19, 20],
[11, 12, 13, 14, 15, 16, 17, 2, 18, 19, 20, 21], [11,
12, 13, 14, 15, 16, 17, 2, 18, 19, 20, 21, 22], [11,
12, 13, 14, 15, 16, 17, 2, 18, 19, 20, 21, 22, 23]]

...

5. Pad the sequence. Since each n-gram sequence is
    different in length, the matrix computation in the
    neural network would not be possible. For this reason,
    each n-gram sequence is padded with 0 to make it

equal in length. For example, the first sequence in the list [3, 4] is padded as [0 0 0 0 0 0 0 0 0 0 0 0 3 4]. The following is the view of the matrix after padding:

```
max_sequence_len = max([len(x) for x in input_review_
sequences])
input_review_sequences = np.array(pad_sequences(input_
review_sequence,
maxlen=max_sequence_len, padding='pre'))
print(input_review_sequences)
```

```
[[ 0  0  0  0  0  0  0  0  0  0  0  0  3  4]
 [ 0  0  0  0  0  0  0  0  0  0  0  3  4  1]
 [ 0  0  0  0  0  0  0  0  0  0  3  4  1  5]
 [ 0  0  0  0  0  0  0  0  0  3  4  1  5  6]
 [ 0  0  0  0  0  0  0  0  3  4  1  5  6  2]
 [ 0  0  0  0  0  0  0  3  4  1  5  6  2  7]
 [ 0  0  0  0  0  0  3  4  1  5  6  2  7  1]
 [ 0  0  0  0  0  3  4  1  5  6  2  7  1  8]
 [ 0  0  0  0  3  4  1  5  6  2  7  1  8  9]
 [ 0  0  0  3  4  1  5  6  2  7  1  8  9 10]
 [ 0  0  0  0  0  0  0  0  0  0  0  0 11 12]
 [ 0  0  0  0  0  0  0  0  0  0  0 11 12 13]
 [ 0  0  0  0  0  0  0  0  0  0 11 12 13 14]
 [ 0  0  0  0  0  0  0  0  0 11 12 13 14 15]
 [ 0  0  0  0  0  0  0  0 11 12 13 14 15 16]
 [ 0  0  0  0  0  0  0 11 12 13 14 15 16 17]
 [ 0  0  0  0  0  0 11 12 13 14 15 16 17  2]
 [ 0  0  0  0  0 11 12 13 14 15 16 17  2 18]
 [ 0  0  0  0 11 12 13 14 15 16 17  2 18 19]
 [ 0  0  0 11 12 13 14 15 16 17  2 18 19 20]
 [ 0  0 11 12 13 14 15 16 17  2 18 19 20 21]
 [ 0 11 12 13 14 15 16 17  2 18 19 20 21 22]
```

```
[11 12 13 14 15 16 17  2 18 19 20 21 22 23]
[ 0  0  0  0  0  0  0  0  0  0  0  0 24 25]
[ 0  0  0  0  0  0  0  0  0  0  0 24 25 26]
[ 0  0  0  0  0  0  0  0  0  0 24 25 26 27]
[ 0  0  0  0  0  0  0  0  0 24 25 26 27 28]
[ 0  0  0  0  0  0  0  0 24 25 26 27 28 29]
[ 0  0  0  0  0  0  0 24 25 26 27 28 29 30]
[ 0  0  0  0  0  0 24 25 26 27 28 29 30 31]
[ 0  0  0  0  0 24 25 26 27 28 29 30 31  1]
[ 0  0  0  0 24 25 26 27 28 29 30 31  1 32]
[ 0  0  0 24 25 26 27 28 29 30 31  1 32 33]
[ 0  0 24 25 26 27 28 29 30 31  1 32 33 34]]
```

6. Set the last word as the label for each n-gram
    sequence. For example, in the n-gram sequence
    [3,4] corresponding to the words ["chilling," "in"], the
    label is "in." Moreover, in the n-gram sequence [3,4,1]
    corresponding to the words ["chilling," "in," "the"],
    the label is "the." Since the model is for predicting
    the next possible word as part of the text generation
    process, the sequence of predictor and label will
    help the neural network train on which word is more
    likely to occur next following a sequence of words.
    The following code is the label for each of the above
    n-gram sequences in the above matrix; observe that
    it is the last inter in each row of the above matrix:

```
predictors, label = input_review_sequences[:,:-1],
input_review_sequences[:,-1]
print(label)
```

```
[ 4  1  5  6  2  7  1  8  9 10 12 13 14 15 16 17  2 18
 19 20 21 22 23 25 26 27 28 29 30 31  1 32 33 34]
```

7. As a final step in the preprocessing, we convert each label into a one-hot encoded vector to make it feasible for matrix computation in the neural network training. to_categorical() is a method from the keras.utils library. Here is the output:

```
label = ku.to_categorical(label, num_classes=total_words)
print(label)

[[0. 0. 0. ... 0. 0. 0.]
 [0. 1. 0. ... 0. 0. 0.]
 [0. 0. 0. ... 0. 0. 0.]
 ...
 [0. 0. 0. ... 1. 0. 0.]
 [0. 0. 0. ... 0. 1. 0.]
 [0. 0. 0. ... 0. 0. 1.]]
```

Putting all the above preprocessing into a single method, we get the following code:

```
#Tokenization to extract terms or words from a corpus
review_tokenizer = Tokenizer()
def dataset_preparation(review_data):
    corpus = review_data.lower().split("\n")
    review_tokenizer.fit_on_texts(corpus)
    total_words = len(review_tokenizer.word_index) + 1

    #Convert the corpus into a flat dataset
    input_review_sequences = []
    for line in corpus:
        token_list = review_tokenizer.texts_to_
        sequences([line])[0]
        for i in range(1, len(token_list)):
            n_gram_sequence = token_list[:i+1]
            input_review_sequences.append(n_gram_sequence)
```

```
#Pad the sequences
max_sequence_len = max([len(x) for x in input_review_
sequences])
input_review_sequences = np.array(pad_sequences(input_
review_sequences, maxlen=max_sequence_len, padding='pre'))

#Predictor and label data
predictors, label = input_review_sequences[:,:-1],input_
review_sequences[:,-1]
label = ku.to_categorical(label, num_classes=total_words)

return predictors, label, max_sequence_len, total_words
```

Creating an RNN Architecture Using a LSTM Network

As discussed in the introduction, using the predictors and labels generated in the dataset preprocessing step above, we create a model using the following layers:

1. **Embedding**: It is a dense vector representation for each word index. The fixed integers of the predictor are converted into randomly selected dense vectors. For example, [3,4] could be converted into [[0.26, 0.14], [0.2, -0.4]]. The dimension of the dense vector is provided by the second argument, output_dim, to the embedding method in Keras. The first argument to the method is input_dim, which is the total number of words in the review. The argument input_length is set equal to the max sequence length minus 1.

2. **LSTM**: The long short-term memory layer takes
    `units` as the dimensionality of the output space.
    The activation function by default is tanh, and the
    recurrent activation function is a hard sigmoid
    function by default. Other available activation
    functions are softmax, Rectified Linear Unit (ReLU),
    and others. With LSTMs it is recommended to use
    tanh and sigmoid.

3. **Dropout**: RNN networks have a tendency to
    overfit the data. In the `dropout` method in Keras, it
    randomly sets a fraction of input units to 0 based on
    the value in the argument `rate`. In the example, the
    rate is set to 0.1, which means randomly drop 10% of
    the input units.

4. **Dense**: The `dense` method creates a regular densely
    connected neural network. This holds the output
    layer where a `softmax` activation function is applied
    to give values between 0 and 1. The word with a
    value close to 1 is highly probable to be the next
    word in the sequence based on the input predictor.

Finally, using the `fit` method, we train the model. In the `fit` function,
we give predictors, labels, and epochs as the input arguments. Epochs
decide the number of iterations for training. After the predefined epochs,
the training stops. The `compile()` method sets the `loss` function to
`categorical_crossentropy` and the `adam` optimizer is chosen as a learning
algorithm, which is based on the stochastic descent approach. The metric
accuracy is set to observe the improvement in the training accuracy as the
epochs increase.

#RNN model

```
def create_model(predictors, label, max_sequence_len, total_
words):
    input_len = max_sequence_len - 1
    model = Sequential()
    model.add(Embedding(input_dim = total_words, output_dim = 10,
    input_length=input_len))
    model.add(LSTM(150))
    model.add(Dropout(0.1))
    model.add(Dense(total_words, activation='softmax'))
    model.compile(loss='categorical_crossentropy',
    optimizer='adam')
    model.fit(predictors, label, epochs=100, verbose=1)
    return model
```

Defining the Generate Text Method

The following method, using the trained model, predicts the most probable next word. The word with the highest probability is given as an output of the model. Since the input to the model is the sequence of integers from the word indexes, a final mapping to the corresponding word is performed in the for loop in the following code. A sample seed text is used in the prediction to generate the text. We can control the number of words we would like to generate.

```
def generate_text(seed_text, next_words, max_sequence_len, model):
    for j in range(next_words):
        token_list = review_tokenizer.texts_to_sequences([seed_
        text])[0]
        token_list = pad_sequences([token_list], maxlen=
                            max_sequence_len-1, padding='pre')
        predicted = model.predict_classes(token_list, verbose=0)
```

```
        output_word = ""
        for word, index in review_tokenizer.word_index.items():
            if index == predicted:
                output_word = word
                break
        seed_text += " " + output_word
    return seed_text
```

Training the RNN Model

Finally, now we use the `dataset_preparation()` method to prepare the data and then pass the output to the `create_model()` method to start the training. The training automatically stops after 100 epochs. Since epoch is a hyperparameter, we could change the value to reduce the loss value further.

```
X, Y, max_len, total_words = dataset_preparation(review_data)
model = create_model(X, Y, max_len, total_words)

Epoch 1/100
34/34 [==============================] - ETA: 0s - loss:
3.5555 - acc: 0.0000e+0 - 4s 129ms/step - loss: 3.5560 - acc:
0.0000e+00
Epoch 2/100
34/34 [==============================] - ETA: 0s - loss:
3.5527 - acc: 0.093 - 0s 1ms/step - loss: 3.5528 - acc: 0.0882
Epoch 3/100
34/34 [==============================] - ETA: 0s - loss:
3.5514 - acc: 0.093 - 0s 1ms/step - loss: 3.5513 - acc: 0.0882
Epoch 4/100
34/34 [==============================] - ETA: 0s - loss:
3.5492 - acc: 0.187 - 0s 1ms/step - loss: 3.5497 - acc: 0.1765
```

```
...
Epoch 79/100
34/34 [==============================] - ETA: 0s - loss:
2.2720 - acc: 0.312 - 0s 1ms/step - loss: 2.3008 - acc: 0.2941
Epoch 80/100
34/34 [==============================] - ETA: 0s - loss:
2.4143 - acc: 0.250 - 0s 1ms/step - loss: 2.4352 - acc: 0.2647
Epoch 81/100
34/34 [==============================] - ETA: 0s - loss:
2.2882 - acc: 0.187 - 0s 2ms/step - loss: 2.2994 - acc: 0.1765
Epoch 82/100
34/34 [==============================] - ETA: 0s - loss:
2.6602 - acc: 0.187 - 0s 1ms/step - loss: 2.7360 - acc: 0.1765
Epoch 83/100
34/34 [==============================] - ETA: 0s - loss:
2.5597 - acc: 0.250 - 0s 1ms/step - loss: 2.5235 - acc: 0.2353
Epoch 84/100
34/34 [==============================] - ETA: 0s - loss:
2.2769 - acc: 0.218 - 0s 1ms/step - loss: 2.2392 - acc: 0.2353
Epoch 85/100
34/34 [==============================] - ETA: 0s - loss:
2.4094 - acc: 0.218 - 0s 1ms/step - loss: 2.4340 - acc: 0.2059
Epoch 86/100
34/34 [==============================] - ETA: 0s - loss:
2.4646 - acc: 0.187 - 0s 1ms/step - loss: 2.4646 - acc: 0.1765
Epoch 87/100
34/34 [==============================] - ETA: 0s - loss:
2.3705 - acc: 0.218 - 0s 1ms/step - loss: 2.3532 - acc: 0.2353
Epoch 88/100
34/34 [==============================] - ETA: 0s - loss:
2.2616 - acc: 0.312 - 0s 1ms/step - loss: 2.2674 - acc: 0.2941
```

```
Epoch 89/100
34/34 [==============================] - ETA: 0s - loss:
2.3206 - acc: 0.156 - 0s 1ms/step - loss: 2.3513 - acc: 0.1765
Epoch 90/100
34/34 [==============================] - ETA: 0s - loss:
2.3629 - acc: 0.187 - 0s 1ms/step - loss: 2.3760 - acc: 0.2059
Epoch 91/100
34/34 [==============================] - ETA: 0s - loss:
2.3248 - acc: 0.218 - 0s 1ms/step - loss: 2.3491 - acc: 0.2059
Epoch 92/100
34/34 [==============================] - ETA: 0s - loss:
2.1996 - acc: 0.218 - 0s 1ms/step - loss: 2.2334 - acc: 0.2059
Epoch 93/100
34/34 [==============================] - ETA: 0s - loss:
2.2162 - acc: 0.156 - 0s 1ms/step - loss: 2.2047 - acc: 0.1765
Epoch 94/100
34/34 [==============================] - ETA: 0s - loss:
2.2623 - acc: 0.250 - 0s 1ms/step - loss: 2.2318 - acc: 0.2647
Epoch 95/100
34/34 [==============================] - ETA: 0s - loss:
2.3510 - acc: 0.218 - 0s 1ms/step - loss: 2.3256 - acc: 0.2353
Epoch 96/100
34/34 [==============================] - ETA: 0s - loss:
2.3909 - acc: 0.218 - 0s 1ms/step - loss: 2.3408 - acc: 0.2647
Epoch 97/100
34/34 [==============================] - ETA: 0s - loss:
2.1507 - acc: 0.250 - 0s 1ms/step - loss: 2.1700 - acc: 0.2353
Epoch 98/100
34/34 [==============================] - ETA: 0s - loss:
2.2254 - acc: 0.218 - 0s 1ms/step - loss: 2.1525 - acc: 0.2353
Epoch 99/100
```

```
34/34 [==============================] - ETA: 0s - loss:
2.1904 - acc: 0.281 - 0s 1ms/step - loss: 2.1384 - acc: 0.2941
Epoch 100/100
34/34 [==============================] - ETA: 0s - loss:
2.1210 - acc: 0.281 - 0s 1ms/step - loss: 2.1275 - acc: 0.2941
```

Generating Text

Now, using the model, we can predict the next word given a seed text. In the following example, the seed text is "signing like," and we ask to predict the next three words. The results are near what we expect. However, instead for predicting "janet," it predicted "jackson." Note that we took a small sample of data to train the model. More data would further improve performance. As we also observed in training, the training accuracy by the end of 100 epochs stayed at 29%, which is not quite high.

```
text = generate_text("singing like", 3, max_len, model)
print(text)

singing like jackson jackson the
```

Applications

In this section, using the knowledge gained so far, we will build the following four applications of NLP:

- **Topic modeling using the spaCy, NLTK, and gensim libraries:** This is an extension of the topic modeling we performed using LDA earlier in the chapter. In this demonstration, we will use the combined knowledge of spaCy, NLTK, and gensim to perform various tasks in topic modeling.

- **Classify between male and female gender by using the person name:** Using features like the last letter of a name and a corpus of male and female names, we will classify between a male and female name. This might help in filtering through the reviews and identifying any gender-based distinctions in the reviews for a product.

- **Given a document, classifying it into a different category:** Classify a review into positive and negative. We will use the NLTK library to perform the preprocessing and classification using the Naïve Bayes classifier.

- **Intent classification and question answering:** In this application, we will build an intent classifier and context-based question-answering utility which could be integrated with any chatbot application. We will use pretrained deep learning models using the DeepPavLov library in Python.

Topic Modeling Using spaCy, NLTK, and gensim Libraries

In the demonstration, we will use spaCy for tokenizing the review text, NLTK for the lemmatization and preprocessing the text, and the LDA model from gensim for training the model.

Tokenizing and Cleaning the Text

Using the en_core_web_md language model in spaCy (which is a slightly bigger pretrained model than sm, meaning it's trained on the larger vocabulary of words), we will do the following in the cleaning process for each token:

1. Detect URLs and screen names, and append them separately into the lda_review_tokens list. This is to ensure the URLs and screen names are not processed further.

2. Convert the rest of the tokens into lowercase.

```python
# Clean
import spacy
spacy.load('en_core_web_md')
from spacy.lang.en import English
parser = English()

def tokenize_review_text(text):
    lda_review_tokens = []
    review_tokens = parser(text)
    for token in review_tokens:
        if token.orth_.isspace():
            continue
        elif token.like_url:
            lda_review_tokens.append('URL')
        elif token.orth_.startswith('@'):
            lda_review_tokens.append('SCREEN_NAME')
        else:
            lda_review_tokens.append(token.lower_)
    return lda_review_tokens
```

Lemmatization

Using the wordnext method, return the lemma for each word. Lemmatization keeps only the root of the word, not its different forms.

```
import nltk
nltk.download('wordnet')
from nltk.corpus import wordnet as wordNet

def get_lemma(word):
    lemma = wordNet.morphy(word)
    if lemma is None:
        return word
    else:
        return lemma
```

Preprocessing the Text Method for LDA

In the preprocessing step, we perform the following functions:

1. Remove all the stopwords in the English vocabulary. We need to download the dataset named stopwords before we can check for the presence of them in the token.

2. Extract the lemma for each token after removing the stopwords.

The following code shows the result of preprocessing on a sample text:

```
from nltk.stem.wordnet import WordNetLemmatizer
def get_lemma2(word):
    return WordNetLemmatizer().lemmatize(word)

# Remove English stopwords
nltk.download('stopwords')
```

```
en_stop = set(nltk.corpus.stopwords.words('english'))

def preprocess_text_for_lda(input_review_text):
    tokens = tokenize_review_text(input_review_text)
    tokens = [token for token in tokens if len(token) > 4]
    tokens = [token for token in tokens if token not in en_stop]
    tokens = [get_lemma(token) for token in tokens]
    return tokens

preprocess_text_for_lda("I consume about a jar every two weeks of
this, either adding it to fajitas or using it as a corn chip dip")

['consume', 'every', 'week', 'either', 'add', 'fajitas', 'using']
```

Reading the Training Data

We read the review file named corn_review.txt, which contains a few
sample reviews related to a "corn" based product in the Amazon Fine
Food review dataset. The following code prints the first few reviews after
preprocessing the reviews from the file:

```
review_text_data = []
with open('data/corn_review.txt') as f:
    for line in f:
        tokens = preprocess_text_for_lda(line)
        print(tokens)
        review_text_data.append(tokens)

['consume', 'every', 'week', 'either', 'add', 'fajitas', 'using']
['taste', 'taste', 'check', 'ingredient']
['found', 'crisp', 'local', 'walmart', 'figure', 'would']
...
```

Bag of Words

Now using the gensim library, we convert the processed review text from the previous step into a bag-of-words corpus and store it on the disk as a pickle file. We later load the file and train the LDA model. Also, we save the dictionary of words created using corpora.Dictionary.

#LDA gensim

```
from gensim import corpora
corn_review_dict = corpora.Dictionary(review_text_data)
corn_review_corpus = [corn_review_dict.doc2bow(text) for text
in review_text_data]
```

```
import pickle
pickle.dump(corpus, open('corn_review_corpus.pkl', 'wb'))
dictionary.save('corn_review_dict.gensim')
```

Training and Saving the Model

Finally, using the ldamodel from genism, we train the model to generate five topics and save the model on disk for later use. Observe that the model gives topic representation using words and their weights in deciding the topic.

```
import gensim
number_of_topics = 5
corn_review_ldamodel = gensim.models.ldamodel.LdaModel(corn_
review_corpus, num_topics = number_of_topics, id2word=corn_
review_dict, passes=15)
corn_review_ldamodel.save('corn_review_ldamodel.gensim')
topics = corn_review_ldamodel.print_topics(num_words=4)
for topic in topics:
    print(topic)
```

```
(0, '0.020*"ginger" + 0.018*"flavor" + 0.015*"recipe" +
0.015*"syrup"')
(1, '0.021*"chips" + 0.014*"tortilla" + 0.014*"flavor" +
0.014*"rather"')
(2, '0.016*"using" + 0.016*"add" + 0.016*"fajitas" +
0.016*"consume"')
(3, '0.003*"ginger" + 0.003*"vernor" + 0.003*"taste" +
0.003*"sugar"')
(4, '0.034*"taste" + 0.019*"check" + 0.019*"ingredient" +
0.003*"product"')
```

From the output above, it looks like topics 0 and 3 are about a "ginger flavor corn syrup" while topics 2 and 4 are not very clear on what they convey. Moreover, topic 1 talks about "tortilla chips."

Predictions

Now, using the above model, let's see how well the model does on a new text. In order to predict the topic, we need to first preprocess and convert the corpus into a bag-of-words representation. From the prediction, it looks like the first example is more related to topic 0, which has the highest probability. Moreover, the second example talks about "tortilla chips," which is represented by topic 1 above.

```
#Prediction
new_doc = 'Corn is typically yellow but comes in a variety of
other colors, such as red, orange, purple, blue, white, and
black.'
new_doc = preprocess_text_for_lda(new_doc)
new_doc_bow = corn_review_dict.doc2bow(new_doc)
print(new_doc_bow)
print(corn_review_ldamodel.get_document_topics(new_doc_bow))
```

```
[(100, 1), (219, 1)]
[(0, 0.73304677), (1, 0.066701755), (2, 0.0667417), (3,
0.066757984), (4, 0.066751845)]
```

```
new_doc = 'corn tortilla or just tortilla is a type of thin,
unleavened flatbread'
new_doc = preprocess_text_for_lda(new_doc)
new_doc_bow = corn_review_dict.doc2bow(new_doc)
print(new_doc_bow)
print(corn_review_ldamodel.get_document_topics(new_doc_bow))
```

```
[(230, 2)]
[(0, 0.06699851), (1, 0.73296124), (2, 0.066678636),
(3, 0.06668124), (4, 0.06668032)]
```

Gender Identification

In this application, we use a corpus of male and female names to build a
model for predicting gender from a given name. It is a simple model with
the only feature as the last letter of the name. The core idea is that female
and male names generally show certain distinctive features. For example,
most female names end with a, e, and i. We use the NLTK library to build
this model.

Loading the NLTK Library and Downloading the Names Corpus

Download the male and female name corpus from the NLTK library. The
corpus mostly consists of English names. The model is generic and is
applicable to non-English names. However, note that the feature we derive
might not be applicable for all names.

```
import nltk
nltk.download('names')

[nltk_data] Downloading package names to
[nltk_data]     C:\Users\KARTHIK\AppData\Roaming\nltk_data...
[nltk_data]   Unzipping corpora\names.zip.
```

Loading the Male and Female Names

After downloading, we create a list of male and female separately to process it further.

```
names = nltk.corpus.names
names.fileids()

male_names = names.words('male.txt')
female_names = names.words('female.txt')
```

Common Names

We can print a few common names that are in both the male and female corpus, such as Abbie, Andy, and Barrie.

```
#Common names
print([w for w in male_names if w in female_names])

['Abbey', 'Abbie', 'Abby', 'Addie', 'Adrian', 'Adrien', 'Ajay',
'Alex', 'Alexis', 'Alfie', 'Barrie', 'Ariel', 'Allie', 'Angel'
, 'Angie' , 'Andrea', 'Andy', 'Allyn', 'Andie', 'Alix',
'Ashley', 'Aubrey', 'Augustine', 'Austin', 'Averil', 'Ali',
'Barry', 'Beau', 'Bennie', 'Benny',...]
```

Extract Features

As a feature to our model, we extract the last letter of each name. Generally, the last name is a good indicator of a person's gender. We will further see in the output of the model how the last letter of the person name plays an important role in the gender prediction model.

```
def gender_features(word):
    return {'last_letter': word[-1]}
gender_features('Shrek')

{'last_letter': 'k'}
```

Randomly Splitting into Train and Test

Now we train the model. We split the male and female corpus of names into training and testing sets. The split is chosen after shuffling the names randomly using the library random in Python. From the resulting corpus, we assign the first 500 names into training and the next 500 into testing.

```
from nltk.corpus import names
labeled_names = ([(name, 'male') for name in names.words('male.
txt')] + [(name, 'female') for name in names.words('female.
txt')])

import random
random.shuffle(labeled_names)

featuresets = [(gender_features(n), gender) for (n, gender) in
labeled_names]
train_set, test_set = featuresets[500:], featuresets[:500]
```

Training the Model

We train the model using the Naïve Bayes (NB) classifier on the training dataset. NB is based on the Bayes Theorem, which computes the prior and posterior probabilities based on whether a given name is male or female. The discussion on NB is beyond the scope of this book. Interested readers can refer to the NLTK documentation at the following link which explains the implementation: `www.nltk.org/_modules/nltk/classify/naivebayes.html`.

```
classifier = nltk.NaiveBayesClassifier.train(train_set)
```

Model Prediction

Using the model built above, we predict the gender of a few names like John and Sascha. Also, we try a few common names and see in which class the model predicts.

```
classifier.classify(gender_features('John'))
```

```
'male'
```

```
classifier.classify(gender_features('Sascha'))
```

```
'female'
```

Model Accuracy

The model seems to have an accuracy of 81.6%, which is quite a good model. We need to incorporate more features if we wish to be more precise in the prediction.

```
print(nltk.classify.accuracy(classifier, test_set))
```

```
0.816
```

Most Informative Features

Using the show_most_informative_features() method from the model, we can see which last letters from the names are essential for classifying the male and female names.

Looking the following output, a name that contains **a** as the last letter is almost 36 times more likely to be female than male, while a name that has **k** as the last letter is 32 times more likely to be male. The accuracy of this model is more than 80%.

```
classifier.show_most_informative_features(5)
```

```
Most Informative Features
        last_letter = 'a'        female : male =     35.7 : 1.0
        last_letter = 'k'        male : female =     32.0 : 1.0
        last_letter = 'p'        male : female =     19.7 : 1.0
        last_letter = 'f'        male : female =     15.8 : 1.0
        last_letter = 'v'        male : female =      9.8 : 1.0
```

Document Classification

A common task in NLP is when we tag a document (could also be a collection of sentences) into a specific category. An example is a news aggregator classifying articles into political, sports, and business. Such classification is useful when there is an enormous amount of unstructured textual data, and no manual labor is available for tagging them. The automatic document classifier could speed-track the process of tagging. Another domain where it's useful is in classifying movie and product reviews into positive and negative sentiments.

Loading Libraries

We will use the `CategorizedPlaintextCorpusReader` method from the NLTK library to create a corpus of review with categories stored with it.

```
import os
import random
from nltk.corpus.reader.plaintext import
CategorizedPlaintextCorpusReader
```

Reading the Dataset into the Categorized Corpus

We have created two sets of reviews, negative and positive. Each positive and negative review is stored in a separate text file with names like 1_neg.txt and 1_pos.txt, and put into a common folder. The following code reads each of the files and categorizes the review into either "pos" for positive and "neg" for negative. There are 10 text files in each of the categories. This is stored as `CategorizedPlaintextCorpusReader`.

```
# Directory of the corpus
corpusdir = 'corpus/'
review_corpus = CategorizedPlaintextCorpusReader(corpusdir,
r'.*\.txt', cat_pattern=r'\d+_(\w+)\.txt')

# list of documents(fileid) and category (pos/neg)
documents = [(list(review_corpus.words(fileid)), category)
             for category in review_corpus.categories()
             for fileid in review_corpus.fileids(category)]
random.shuffle(documents)

for category in review_corpus.categories():
    print(category)
```

output:
neg

pos

```
type(review_corpus)
```

```
nltk.corpus.reader.plaintext.CategorizedPlaintextCorpusReader
```

```
len(documents)
```

20

Computing Word Frequency

Now we count the frequency of occurrence of each word in a given corpus using the FreqDist() method from NLTK. The following code prints the top 200 words in descending order of frequency of occurrence:

```
import nltk
all_words = nltk.FreqDist(w.lower() for w in review_corpus.words())
word_features = list(all_words)[:200]
```

```
print(word_features)
```

```
['warning', '!', '-', 'alcohol', 'sugars', '!,"', 'buyer',
'beware', 'please', 'this', 'sweetener', 'is', 'not',
'for', 'everybody', '.', 'maltitol', 'an', 'sugar', 'and',
'can', 'be', 'undigestible', 'in', 'the', 'body', 'you',
'will', 'know', 'a', 'short', 'time', 'after', 'consuming',
'it', 'if', 'are', 'one', 'of', 'unsuspecting', 'many',
'who', 'cannot', 'digest', 'by', 'extreme', 'intestinal',
'bloating', 'cramping', 'massive', 'amounts', 'gas', 'person',
'experience', 'nausea', ',', 'diarrhea', '&', 'headaches',
'also', 'experienced', 'i', 'learned', 'my', 'lesson', 'hard',
'way', 'years', 'ago', 'when', 'fell', 'love', 'with', 'free',
'chocolates', 'suzanne', 'sommers', 'used', 'to', 'sell',
'thought', "'", 'd', 'found', 'chocolate', 'nirvana', 'at',
'first', 'taste', 'but', 'bliss', 'was',..]
```

Checking the Presence of Frequent Words

We define a method called document_features(), which checks whether a frequent word is present in any of the neg and pos review text files read earlier. If it finds a frequent contains, the print statement will print the word.

```
#Check whether most frequent word is present in the doc or not
def document_features(document):
    document_words = set(document)
    features = {}
    for word in word_features:
        features['contains({})'.format(word)] = (word in
        document_words)
    return features

print(document_features(review_corpus.words('1_pos.txt')))

{'contains(warning)': False, 'contains(!)': False,
'contains(-)': False, 'contains(alcohol)': False,
'contains(sugars)': False, 'contains(!,")': False,
'contains(buyer)': False,...}

print(document_features(review_corpus.words('1_neg.txt')))

{'contains(warning)': False, 'contains(!)': False,
'contains(-)': False, 'contains(alcohol)': False,
'contains(sugars)': False, 'contains(!,")': False,...}
```

Training the Model

We use 15 randomly selected docs for training and 5 for testing. We then use the Naïve Bayes classifier for classification. We also print the accuracy on testing and training data. It seems to give a very low accuracy of 20% on testing and 67% on training. The accuracy could be improved with more data training data.

```
featuresets = [(document_features(d), c) for (d,c) in documents]
train_set, test_set = featuresets[5:], featuresets[:5]
classifier = nltk.NaiveBayesClassifier.train(train_set)

print(nltk.classify.accuracy(classifier, test_set))

0.2

print(nltk.classify.accuracy(classifier, train_set))

0.6666666666666666
```

Most Informative Features

Again, using the show_most_informative_features from the model, we check which words are more likely to decide whether a review will be negative or positive. This gives an explanation for why the review was classified as negative and positive.

```
classifier.show_most_informative_features(5)

Most Informative Features
          contains(not) = True        neg : pos   =      5.2 : 1.0
         contains(this) = False       neg : pos   =      5.2 : 1.0
         contains(like) = True        neg : pos   =      4.3 : 1.0
          contains(not) = False       pos : neg   =      4.0 : 1.0
         contains(this) = True        pos : neg   =      4.0 : 1.0
           contains(so) = True        neg : pos   =      3.3 : 1.0
```

```
contains(me) = True          neg : pos   =    3.3 : 1.0
contains(good) = True        neg : pos   =    2.6 : 1.0
contains(have) = True        neg : pos   =    2.6 : 1.0
contains(much) = True        neg : pos   =    2.4 : 1.0
```

In this corpus, a review that mentions **"not"** is almost five times more likely to be negative than positive, while a review that mentions **"good"** is only about three times more likely to be negative than positive. Perhaps the negative-ness of the word "good" might stem from the customers with reviews of the nature, "the product is good but …" where they may have one or two complain.

If we add more reviews to this corpus of positive and negative, the accuracy will start to improve.

Intent Classification and Question Answering

The two most important NLU tasks a chatbot should perform well are to classify the intent of a given user query and answer questions by understanding the context. While there are many propriety frameworks around these two tasks, they don't provide the visibility of what happens behind the scene. In this section, we will use a Python library called deeppavlov. It's an open-source deep learning library for end-to-end dialog systems and chatbots. The library provides many pretrained deep learning models as part of its offering.

Intent Classification

We need to classify a given query (input from the user) into an intent class. Once an intent class is identified, a chatbot can trigger the respective logic as a response to a user query. For example, if the query is "how is the weather today," the intent classification should trigger the weather services API from within the chatbot and fetch the result.

The deeppovlav library provides many built-in intent classification models. In the following demo, we will use a pretrained NLU benchmark dataset called SNIPS. It is trained for the following seven intents:

- GetWeather

- BookRestaurant

- PlayMusic

- AddToPlaylist

- RateBook

- SearchScreeningEvent

- SearchCreativeWork

Setting tensorflow as the Back End

In order to use the KerasClassificationModel in the Windows platform, we need to set the KERAS_BACKEND to "tensorflow". The following code is used for the same:

```
import os
os.environ["KERAS_BACKEND"] = "tensorflow"
```

Building the Model

We install deeppavlov in either the virtualenv or conda environments. In the following command line example, we create a conda environment named deeppavlov and then install and download the required libraries and model files for using SNIPS intents:

```
(deeppavlov) C:\Users\Karthik\ Code>python -m deeppavlov
install "C:\ProgramData\Anaconda3\Lib\site-packages\deeppavlov\
configs\classifiers\intents_snips.json"
```

```
(deeppavlov) C:\Users\Karthik\ Code>python -m deeppavlov
download "C:\ProgramData\Anaconda3\Lib\site-packages\
deeppavlov\configs\classifiers\intents_snips.json"
```

Once the installation and download is successful, the following code builds the model using build_model method. Note that the first time you run this code, you need to set the download = True for downloading all required pretrained model. The size of download is approximately 3GB.

```
from deeppavlov import build_model, configs
CONFIG_PATH = configs.classifiers.intents_snips  # could also
be configuration dictionary or string path or `pathlib.Path`
instance
#model = build_model(CONFIG_PATH, download=True)  # run it once
model = build_model(CONFIG_PATH, download=False)  # otherwise
```

```
2019-07-02 19:48:10.74 INFO in 'deeppavlov.models.embedders.
fasttext_embedder'['fasttext_embedder'] at line 67: [loading
fastText embeddings from `C:\Users\Karthik\.deeppavlov\
downloads\embeddings\dstc2_fastText_model.bin`]
Using TensorFlow backend.
2019-07-02 19:51:04.703 INFO in 'deeppavlov.models.classifiers.
keras_classification_model'['keras_classification_model'] at
line 273: [initializing `KerasClassificationModel` from saved]
2019-07-02 19:51:05.866 INFO in 'deeppavlov.models.classifiers.
keras_classification_model'['keras_classification_model'] at
line 283: [loading weights from model.h5]
2019-07-02 19:51:07.653 INFO in 'deeppavlov.models.classifiers.
keras_classification_model'['keras_classification_model'] at
line 134: Model was successfully initialized!
Model Summary:

...
```

```
Total params: 235,475
Trainable params: 233,725
Non-trainable params: 1,750
```

Classifying the Intent

Now we can use the model. In the following code, we try a few intents like GetWeather, BookRestaurant, RateBook, and SearchScreeningEvent.

```
print(model(["will it rain in Edgbaston, Birmingham today?"]))
```

```
[['GetWeather']]
```

```
print(model(["book one table at a good restaurant?"]))
```

```
[['BookRestaurant']]
```

```
print(model(["Give Da Vinci Code a 5 star on my amazon purchase"]))
```

```
[['RateBook']]
```

```
print(model(["what are the show times for The Lion King"]))
```

```
[['SearchScreeningEvent']]
```

You can train a custom model to classify the intent for a specific use case. More details on training a custom model can be found at http://docs.deeppavlov.ai/en/latest/components/classifiers. html#how-to-train-on-other-datasets. Training a custom model is a resource-intensive process. So, if you are trying to build a generic chatbot, we suggest you first explore all the pretrained models shown here before deciding to build your own model: http://docs.deeppavlov.ai/en/ latest/components/classifiers.html#pre-trained-models.

Question Answering

Chatbots often need to understand the context of the conversation to answer a particular query from a user. The deeppavlov library provides a pretrained model trained on Stanford Question Answering Dataset (SQuAD) dataset, a reading comprehension dataset consisting of crowdsourced questions on a set of Wikipedia articles. More details on the dataset can be found at `https://rajpurkar.github.io/SQuAD-explorer/`.

The main task the model trained on SQuAD dataset performs is to identify a given context and answer a question within the given context.

Building the Model

Similar to intent classification, we use the `build_model` method with configurations of the SQuAD pretrained model. Run the following code once with `download = True` to get all the required models. Also, run the following command to install the squad_bert pretrained model:

```
python -m deeppavlov install squad_bert
```

```
from deeppavlov import build_model, configs

#model = build_model(configs.squad.squad, download=True)
model = build_model(configs.squad.squad)
```

Context and Question

Now that the model is built, here are some examples of a given context and a question. Then we will see how well the model does. In the first example, we give a context about a chatbot called IRIS, and then ask the model "What is IRIS?" It correctly picks up the most relevant part from the context and gives us the output, starting from the eighth character.

```
model(['IRIS is an enterprise chatbot completely built in-house
and uses private data'], ['What is IRIS?'])
```

```
[['an enterprise chatbot completely built in-house'], [8],
[832987.875]]
```

In the next example, we give one of the reviews from the Amazon Food Review dataset as a context and then ask, "How many cakes were made?" The model is able to give the correct answer as 20.

```
model(['Great morning cake!,We must have made about 20 of these
cakes last fall They are so good. Also very easy to make.  This
was great with bacon and eggs in the morning. It was also great
for dessert (as I believe it was intended ;). We didnt put the
icing on as suggested as the cake was great without it.  Now that
its getting a little chilly out we are excited to start making
our favorite fall cake again'], ['how many cakes were made?'])
```

```
[['20'], [44], [42414.3046875]]
```

In the following example, we test whether the model is able to identify a phrase in the given context which might tell "Is the customer happy about the purchase?" The model picks up the right phrase, which talks about a particular sentiment: "disappointed." It tells us the customer was happy about the purchase. In this question we haven't used any words from the context, but the model still was able to extract the most appropriate phrase.

```
model(['I used these rainbow jimmies for a rainbow cupcake
topper and added them to rice krispie treats for my daughters
6th birthday.  Obviously, it was a rainbow party.  The package
didnt look like the picture, but I was not disappointed in the
product.  I would buy from this company again.'], ['is the
customer happy about the purchase?'])
```

```
[['I was not disappointed'], [209], [2021.420166015625]]
```

Serving the DeepPavlov Model

In DeepPavlov terminology, each skill or component can be made available as a REST API. Once a skill or component is hosted as a skill, any application or service can call the API to get a response. In the following example, if we host the "intents_snips" component using the following command line argument

```
(deeppavlov) C:\Users\Karthik\ Code>python -m deeppavlov
riseapi "C:\ProgramData\Anaconda3\Lib\site-packages\deeppavlov\
configs\classifiers\intents_snips.json"
```

by the end of the above command, we should see the following output, where a Flask app is created and the API is running on the local host. You can specify your own port and URL for hosting the API. More on this can be found at http://docs.deeppavlov.ai/en/latest/devguides/rest_api.html.

```
* Serving Flask app "deeppavlov.utils.server.server" (lazy loading)
* Environment: production

  Use a production WSGI server instead.
* Debug mode: off
* Running on http://0.0.0.0:5000/ (Press CTRL+C to quit)
```

Now, a POST request like the following should return a JSON response with the intent class [['SearchScreeningEvent']]:

```
{"context":[" what are the show times for The Lion King"]}
```

In the next chapter, we will introduce our enterprise chatbot named IRIS, where we can directly call the above REST API for intent classification. Note that you still have to train your own model on the private enterprise data in order to integrate it with the chatbot. Even though we will build IRIS using a Java framework, the REST API we have created above is easily called from within the Java application. We can create many applications using the powerful libraries of Python for NLP, NLU, and NLG tasks and simply host all of it as a REST API, which is language and platform agnostic.

Summary

We started by identifying the differences between natural language processing, understanding, and generation, and then discussed various open source tools available to process and understand natural languages.

Then we delved into NLP, where we showed how to use tools like NLTK, spaCy, CoreNLP, genism, and TextBlob for various task such as processing textual data, normalizing text, part-of-speech tagging, dependency parsing, spelling correction, machine translation, and named entity recognition.

In the NLU section, we showed language models like Word2Vec and GloVe for performing out-of-the-box tasks such as word and sentence similarity, finding linear substructures between words, and performing arithmetic operations on word embedding vectors to find meaningful semantic relationships between words. As an important part of NLG, we explored the relationship extraction from a given sentence using the OpenIE tool and built a topic modelling tool using latent Dirichlet allocation (LDA).

We then moved into NLG, where we explored use cases like a random headline generator using the markovify library in Python. And then we explored SimpleNLG, an English grammar-based natural language generation utility. It offers grammatical structures such as generating the past tense, negation, complements, and prepositional phrases. In the NLG section, we built a deep learning-based model for predicting the next word in a given phrase or a sentence. The model used a popular deep learning architecture called long short-term memory.

In the final part, we covered applications of NLP and NLU: topic modelling, gender, document classification, intent classification, and question answering. In the topic modeling, we utilized all of the available open source tools from the previous sections of the chapter.

Overall, in this chapter we explored extensively the P-U-G of natural languages. The availability of many open source tools from Python and Java facilitated a great number of demonstrations to understand and model natural languages. We covered a varied level of topics starting from parsing text data to building generative models using a deep learning model. Our aim with this chapter was to provide an exhaustive collection of methods and tools to empower you to build chatbots with basic and advanced levels of natural language processing, understanding, and generation capabilities.

Next, we will build and deploy a fully functional in-house enterprise chatbot on private datasets. Since there are many chatbot frameworks with support for NLP and NLU, the methods discussed in this chapter at first might seem not so readily usable; however, under the hood, many frameworks like RASA and LUIS internally uses the techniques discussed in this chapter. Also, many ideas from NLG are still not available in any standard chatbot framework, so they are often built from scratch. We believe the ideas taught in this chapter will come handy when you build an enterprise chatbot.

## CHAPTER 6

A Novel In-House Implementation of a Chatbot Framework

In previous chapters, we explained intents and different ways of classifying intents using natural language techniques. We also discussed the various data sources that are available in designing an enterprise chatbot. There are many chatbot builder platforms and frameworks available in the market that can be used to build chatbots. These frameworks abstract much complex functionality and provide components that are reusable, extendable, and scalable.

Designing an enterprise chatbot without using a framework has the following benefits:

- Provides better security and control

- Data protection from third-party vendors

- Minimizes operational cost

- In-depth analytics

© Abhishek Singh, Karthik Ramasubramanian, Shrey Shivam 2019
A. Singh et al., *Building an Enterprise Chatbot*,
https://doi.org/10.1007/978-1-4842-5034-1_6

- Flexible design of architecture

- Change control management

- Interoperability

- Easy and quick integration with enterprise-wide available services and frameworks

- Integration with messenger platforms

- Integration with custom machine learning models

- Flexibility to customize with changes in the organization

In this chapter, we will discuss and implement a custom-built chatbot called IRIS (Intent Recognition and Information Service). We will explain the implementation concepts of the understanding developed in previous chapters. We will talk about designing and implementing state machines, transitions from one state to another in a conversational chatbot, and how they are critical to maintaining the context of user utterances as well as how to make a chatbot mimic human conversation with short-term memory and long-term memory.

In IRIS, the backbone engine of the chatbot is written in Java, and an integration module connects IRIS with different messenger platforms such as Facebook Messenger. The integration module is written in NodeJS, which is discussed in the next chapter.

Introduction to IRIS

We developed IRIS as an open source chatbot framework to provide a novice level of understanding and implementation of a chatbot from scratch. IRIS provides the ability to use our templates of machine learning-based extraction of information from users' utterances, such as by using named-entity recognition (NER). It provides customized enhancements such as custom intent matching implementation and conversational state management and many other features.

The design is inspired by our collective experience and exploration of how other popular frameworks such as Amazon's Echo natural language understanding model, Alexa Skills, RASA, Mutters, Dialogflow, and Microsoft Bot Builder are designed and implemented. IRIS derives many methods and implmentation from Mutters, an open source Java-based framework for building bot brains, and reuses some of its design and code concepts to create a simple and modified backend code base. Platforms like Mutters provide a lot of out-of-box features, support, and easy integration. Apart from our custom chatbot framework, we will discuss widely popular platforms and frameworks, and how they work, in the next chapter.

Intents, Slots, and Matchers

In the previous chapter, we showed that intents are an outcome of the behavior and focus of the user's utterance. In this chapter, we will describe various components of an intent and discuss the implementation approach of creating and classifying intents. The following components define an intent:

- A name
- Sample utterances
- Slots (entities)
- A slot matcher

Each intent can have zero or more slots, which are used to extract entities from user utterances. For example, if a chatbot helps us discover restaurants nearby, one of the intents could be defined as the following:

Intent name	Restaurant search
Sample utterances	• Looking for restaurants around me
	• Restaurants nearby
	• Best restaurants near me
	• Good continental restaurants nearby
	• Best Chinese restaurants
Slots/Entity	Cuisine
Slot matcher	Custom Entity Match Model

Figure 6-1 explains the meaning of intent, slot, and utterance.

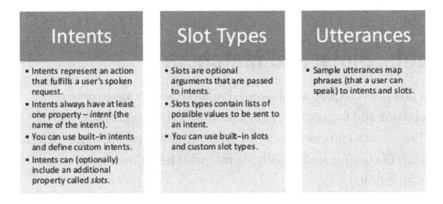

Figure 6-1. *The meanings of intent, slot, and utterance*

Now, we'll procedurally go through all the steps involved in creating intent, slot, and matcher classes in Java.

In our new Java project, we will create a package called `com.iris.bot.intent` in which we will define classes required for intent creation and classification.

Intent Class

We define a Java class called `Intent` with a name variable to store the intent name. Intent has slots that contain a list of 0 or more slots defined for this particular intent. We have getters and setters of the name and slots.

```java
public class Intent {
    /** The name of the intent. */
    protected String name;

    /** The slots for the intent.
     * There could be 0 or more slots defined for each intent.
     * Slots contain a list of Slot and methods to add and
       get Slot  */
    protected Slots slots = new Slots();

    /**
     Constructor with the name as a parameter.
     It sets the Intent name at the time of intent creation.
     */
    public Intent(String name) {
        this.name = name;
    }

    /**
     * Returns the name of the intent.
     */
    public String getName() {
        return name;
    }

    /**
     * Adds a slot to the intent.
     */
```

```java
public void addSlot(Slot slot) {
    slots.add(slot);
}

/**
 * Returns the slots for the intent.
 */
public Collection<Slot> getSlots() {
    return Collections.unmodifiableCollection(slots.
    getSlots());
}
}
```

Now that we have defined Intent, we need to define the IntentMatcherService class.

IntentMatcherService Class

This service takes user utterances and responds with the matched intent. As explained in the previous chapter, there are multiple ways to classify intents. In this example, we have a separate intent classification service that classifies user utterances into one of the user-defined intents with a certain probability or score (refer to Chapter 5 for more details on intent classification).

```java
public class IntentMatcherService {

    /** A map of possible intent names and intents that are
        defined in the Iris Configuration */
    private HashMap<String, Intent> intents = new
    HashMap<String, Intent>();
```

```java
/** The slot matcher method to use for named entity
    recognition. */
private CustomSlotMatcher slotMatcher;

/** Intent Matcher Service constructor that sets slot
    matcher */
public IntentMatcherService(CustomSlotMatcher slotMatcher) {
    this.slotMatcher = slotMatcher;
}

/*
 * RestTemplate is a synchronous Java client to perform
   HTTP requests, exposing a simple template method
   API over underlying HTTP client libraries. The
   RestTemplate offers templates for common scenarios by
   HTTP method, in addition to the generalized exchange
   and execute methods that support less frequent cases.
 */
protected RestTemplate restTemplate = new RestTemplate();

/** This method takes a user utterance and session as
    an input, obtains matched intent from an intent
    classification service, performs named entity
    recognition on slots defined for the matched intent,
    and sets the matched intent into the user session.
    The session is a server-side storage mechanism
    that stores a user's interaction and resets the
    information or persists based on the interaction
    duration and the type of information.
 */
public MatchedIntent match(String utterance, Session
session) {
```

```java
// getIntent method returns the matched intent.
Intent matchedIntent = getIntent(utterance);

/*
 * We define slots associated with each Intent in
   the Iris Configuration class. Each of these
   slots has a matching method defined to describe
   how entities are to be matched. Depending on
   the entity and implementation, various NER
   models can be used to recognize entities. This
   method returns a map of the slot and matched
   slot object. The slot contains a slot name and
   a matching method, and MatchedSlot contains slot
   that was matched, the value that was used to
   match on, and the value that was matched.
 */
HashMap<Slot, MatchedSlot> matchedSlots =
slotMatcher.match(session, matchedIntent, utterance);

/*
 * Once we get the matched intent, we set the
   value of the intent in session. We will discuss
   session under the IRIS Memory topic.
 */
session.setAttribute("currentIntentName",
matchedIntent.getName());

/*
 * Finally, an object with matched intent, matched
   slots, and the utterance against which the
   intents and slots were matched and returned.
 */
```

```java
        return new MatchedIntent(matchedIntent,
        matchedSlots, utterance);
    }
}
```

In the code snippet above, `Intent matchedIntent = getIntent(utterance)` is the method that provides intent classification.

The getIntent Method of the IntentMatcherService class

As discussed, there are many ways in which this method can be implemented. It takes a user utterance as an input and returns an `Intent` that is classified with maximum probability by the intent engine. For now, let's see how to define this method in a simple way:

```java
/*
* getIntent method takes a user utterance and returns an object
of type Intent. This is then used by the match method to match
slots for that intent.
    */
    public Intent getIntent(String utterance) {
        /*
        * Intent Response is a plain Java object with
            three attributes - utterance, intent name, and
            probability returned by the intent service.
        */
        IntentResponse matchedIntent = new IntentResponse();
        /*
        * If the intent classification engine is not able
            to classify the utterance into some intent with
```

```
   some threshold or if the engine is unable to
   return a valid response, we fallback it to be a
   general query intent to be on the safe side.   '
 */
String defaultIntentName = "generalQueryIntent";
String matchedIntentName = null;

/*
 * ObjectMapper provides functionality for reading
   and writing JSON, either to and from basic
   POJOs (Plain Old Java Objects), or to and from
   a general-purpose JSON Tree Model (JsonNode),
   as well as related functionality for performing
   conversions. ObjectMapper is a part of the com.
   fasterxml.jackson.databind package, which is a
   high-performance JSON processor for Java.
 */
ObjectMapper mapper = new ObjectMapper();

/*
 * There is specific enumeration that defines
   simple on/off features to set for ObjectMapper.
   ACCEPT_CASE_INSENSITIVE_PROPERTIES is a feature
   that allows for more forgiving deserialization
   of incoming JSON.
   FAIL_ON_UNKNOWN_PROPERTIES is a feature that
   determines whether encountering of unknown
   properties (ones that do not map to a property,
   and no "any setter" or handler can handle it)
   should fail (by throwing a JsonMappingException)
   or not.
 */
```

```
mapper.configure(MapperFeature.ACCEPT_CASE_
INSENSITIVE_PROPERTIES, true)
.configure(DeserializationFeature.FAIL_ON_UNKNOWN_
PROPERTIES, false);

try {
        matchedIntent = restTemplate.
        getForObject("http://localhost:8080" + "/
        intent/" + utterance, IntentResponse.class);

        if(matchedIntent != null && matchedIntent.
        getIntent()!=null){
                matchedIntentName = matchedIntent.
                getIntent();
        }
        else
// If matched intent is null, we consider the default intent to
be the matched intent.
                matchedIntentName = defaultIntentName;
        } catch (Exception e) {
// In case of an exception too, we consider default.
        matchedIntentName = defaultIntentName;
        }

// Finally, we return the intent object with the matched intent
name back to the match method of Intent Matcher Service.
        return intents.get(matchedIntentName);
    }
```

There are two essential things to be discussed in the getIntent method, and we cover them in the next sections.

Intent Classification Service

We are assuming here that there is an intent classification service running on localhost on port 8080 that accepts HTTP GET requests and returns a JSON response:

```
http://localhost:8080/intent/user-utterance
```

Here's the JSON representation of a response:

```
{
"utterance": "i want a life insurance quote",
"intent": "QUOTE",
"probability": 89.5,
}
```

General Query Intent

Most chatbots today are based on general queries and look like an automated Q&A system. The reason for this is that most developers are unsure how to model the chatbot to be conversational. Also, they find it difficult to make the bot interactive.

A general query is never an explicit intent in a chatbot that is conversational and that mimics human conversation modeled as dialogs. Hence, when no intent is matched by the classification engine or if the match probability is not good enough for that utterance, we tend to classify it as general intent. We have seen that this approach is very efficient in practical situations. In another way, if the intent engine is not able to classify the utterance, the utterance could be a generic ask and not aimed for a specific action. We will show later how to use this intent for first looking for an answer in a FAQ repository and then later as a fallback, performing a general search to return a relevant response if possible.

Matched Intent Class

The last thing that we need to include in the `com.iris.bot.intent` package is a `MatchedIntent` class. It holds the intent that was matched, a map of slots that were matched against the defined slots for the intent, and the utterance against which they were matched.

```java
public class MatchedIntent {

    /** The intent that was matched. */
    private Intent intent;

    /** Map of slots that were matched. */
    private HashMap<Slot, MatchedSlot> slotMatches;

    /** The utterance that was matched against. */
    private String utterance;

    /**
     * Constructor.
     *
     * @param intent
     *            The intent that was matched.
     * @param slotMatches
     *            The slots that were matched.
     * @param utterance
     *            The utterance that was matched against.
     */
    public MatchedIntent(Intent intent, HashMap<Slot,
    MatchedSlot> slotMatches, String utterance) {
        this.intent = intent;
        this.slotMatches = slotMatches;
        this.utterance = utterance;
    }
```

```java
/**
 * Returns the Intent that was matched.
 */
public Intent getIntent() {
    return intent;
}

/**
 * Returns the slots that were matched.
 */
public Map<Slot, MatchedSlot> getSlotMatches() {
    return Collections.unmodifiableMap(slotMatches);
}

/**
 * Returns the specified slot match if the slot was
   matched.
 *
 * @param slotName
 *            The name of the slot to return.
 * @return The slot match or null if the slot was not
   matched.
 */
public MatchedSlot getSlotMatch(String slotName) {
    for (MatchedSlot match : slotMatches.values()) {
        if (match.getSlot().getName().
        equalsIgnoreCase(slotName)) {
            return match;
        }
    }
    return null;
}
```

```java
/**
 * Returns the utterance that was matched against.
 *
 * @return The utterance that was matched against.
 */
public String getUtterance() {
        return utterance;
}
}
```

Slot Class

We covered intent, intent matcher service, and matched intents so far. Designing classes of slots is similar to intents. We define slot-related classes in the com.iris.bot.slot package:

```java
/*
 * Slot is defined as an abstract class. The concrete class
   of Slot implements a match method that contain the entity
   recognition logic.
   getName returns the slot name that is described in concrete
   slot classes.
 */
public abstract class Slot {
        public abstract MatchedSlot match(String utteranceToken);
        public abstract String getName();
}
```

With Slot defined, we create slots for the intent. Slots is an attribute specified in the Intent class and slot details are provided in the IRIS configuration.

```java
/** The slots for the intent.
 * There could be 0 or more slots defined for each intent.
 * Slots contain a list of Slot and methods to add and get
   Slot */
public class Slots {

/** The map of slots. */
    private HashMap<String, Slot> slots = new HashMap<String,
    Slot>();

    /**
     * Adds a slot to the map.
     */
    public void add(Slot slot) {
        slots.put(slot.getName().toLowerCase(), slot);
    }

    /**
     * Gets the specified slot from the map.
     */
    public Slot getSlot(String name) {
        return slots.get(name.toLowerCase());
    }

    /**
     * Returns the slots in the map.
     */
    public Collection<Slot> getSlots() {
        return Collections.unmodifiableCollection(slots.
        values());
    }
}
```

Similarly to MatchedIntent class, we create the MatchedSlot class to hold the details of the slots that were matched:

```
/*
 * MatchedSlot contains slot-related information such as the
   slot that was matched, the original value that was was used
   to match on, and the value that was matched.
 */
public class MatchedSlot {
      /** The slot that was matched. */
      private Slot slot;

      /** The original value that was used to match on. */
      private String originalValue;

      /** The value that was matched. */
      private Object matched value;

      public MatchedSlot(Slot slot, String originalValue,
      Object value) {
            this.slot = slot;
            this.originalValue = originalValue;
            this.setMatchedValue(value);
      }

      /**
       * Returns the slot that was matched.
       */
      public Slot getSlot() {
            return slot;
      }

      /** Returns the original value.
       */
```

```java
    public String getOriginalValue() {
        return originalValue;
    }

    /*
     * Returns the matched value.
     */
    public Object getMatchedValue() {
        return matched value;
    }

    /*
     * Sets the matched value in the constructor. The method
       is declared as private as the value is only set in
       constructor.
     */
    private void setMatchedValue(Object matchedValue) {
        this.matchedValue = matchedValue;
    }
}
```

We defined Slot, Slots, and MatchedSlot so far under the com.iris. bot.slot package. Now let's see how a custom slot matcher is defined. CustomSlotMatcher is invoked in IntentMatcherService to get the slot match information once an intent is obtained from the intent classification service:

```java
/*
 * The CustomSlotMatcher class is used to iterate on all the
   slots for the matched intent and execute a match method
   of each of those slots to return all the matched slots.
   This class can be further customized and designed to have
   multiple types of slot matcher implementation.
 */
```

```java
public class CustomSlotMatcher {
    /*
     * The match method takes session, intent, and user
       utterance as an input and returns a map of Slot and
       MatchedSlot details.
     * This method can further contain business logic
       depending upon the implementation.
     */
    public HashMap<Slot, MatchedSlot> match(Session session,
            Intent intent, String utterance) {
        HashMap<Slot, MatchedSlot> matchedSlots = new
        HashMap<Slot, MatchedSlot>();

// Iterate intent to get all slots defined for this matched intent.
        for (Slot slot : intent.getSlots()) {
/*
 * Use case-specific business logic handling.
   askQuoteLastQuestion is a session variable, and its logic
   will be explained when we discuss state machines and
   conversation flow management.
 */
            String slotCheck = String.valueOf(session.
            getAttribute("askQuoteLastQuestion"));
            if (slot.getName().equalsIgnoreCase(slotCheck)
            || slotCheck.equalsIgnoreCase("null")) {

// The match method defined in each slot is executed, and
MatchedSlot is returned.
                MatchedSlot match = slot.
                match(utterance);
```

```
                        if (match != null && match.
                        getMatchedValue() != "null") {
                                matchedSlots.put(slot, match);
                        }
                }
        }
        return matchedSlots;
    }
}
```

We have so far covered how intent- and slot-related classes can be defined for IRIS. We briefly discussed how IRIS memory is managed by session attributes. Let's go through this in some more detail.

IRIS Memory

A conversational chatbot needs to hold certain information to be able to closely mimic human-like responses. IRIS is designed to hold information in memory through sessions.

Long- and Short-Term Sessions

Session contains two types of attributes:

- Long-term attributes
- Short-term attributes

Long-Term Attributes

Certain entities such as name, date of birth, and gender of the user are information that does not change over time. Also, in the real world, we don't expect our advisors and agents to ask these details every time we interact with them. In the current design of IRIS, what we have demonstrated is that not all attributes are reset after the user session. The long-term attributes that span sessions are supposed to be held in a fast, reliable, and persistent storage databases such as Redis. Redis is an in-memory database. In the code snippet that follows, we show this using a HashMap. Information in HashMaps are stored in JVM when the application is running and get cleared when the application goes down. Hence, even though they are long-term attributes, unless we persist them in a permanent storage like SQL databases, we can't retrieve them again.

Short-Term Attributes

Unlike name and gender, certain attributes are limited to the scope of the user session. In most cases, the expectation is that the values will vary in each session. An example is a user providing a ZIP code when asking for an insurance agent nearby or providing a face amount for a life insurance eligibility quotation. Moreover, to manage the conversation flow, certain values such as current intent, state, and last question asked are stored as short-term attributes. Short-term attributes reset with each new session or if a session expires.

The Session Class

The Session class helps in conversation flow management by storing state- and intent-related information in attributes. It also helps in maintaining the information exchange between the user and the server by serving as a temporary storage layer. There is a reset method that reinitialize attributes when called. A session is created with a current timestamp and an empty attributes map.

```java
public class Session {
    /*
     * We defined a session to be 30 minutes long and this is
       number should vary based on use case, and for how long
       do you want a session to be active.
     */
    public long expiryTimeinMilliSec = 30 * 60 * 1000l;

    private HashMap<String, Object> attributes = new
    HashMap<String, Object>();

    /*
     * Long-term attributes do not get reset when the session
       expires or when the reset method is called.
     */
    private HashMap<String, Object> longTermAttributes = new
    HashMap<String, Object>();

    // Time in milliseconds when the session was created.
       This is used to check whether the session is valid.
    private long timestamp;

    /*
     * A default session constructor is called and it assigns
       current time in milliseconds to the timestamp variable.
     */
    public Session() {
        this.timestamp = System.currentTimeMillis();
    }

    public void updateCurrentState(State currentState) {
        attributes.put("current_state", currentState);
    }
```

```java
public void updateCurrentIntent(String currentIntent) {
    attributes.put("current_intent", currentIntent);
}

/*
 * Checks if this is a valid session. Returns boolean.
 */
public boolean isValid() {
    if (timestamp + expiryTimeinMilliSec < System.
    currentTimeMillis())
            return false;
    return true;
}

/*
 * Returns a session attribute.
 */
public Object getAttribute(String attribute) {
    return attributes.get(attribute);
}

/*
 * Sets a session attribute.
 */
public void setAttribute(String key, Object object) {
    attributes.put(key, object);
}

/**
 * Removes the specified attribute from the session.
 */
public void removeAttribute(String attributeName) {
    attributes.remove(attributeName.toLowerCase());
}
```

```
/**
 * Resets the session, removing all attributes. Long-term
   attributes
 * are not removed from the session.
 */
public void reset() {
        attributes = new HashMap<String, Object>();
}
}
```

We also need a helper class called SessionStorage to create sessions as well as maintain Session for each user. Slots matched in the match method are also saved to session for later use.

```
/*
 * Helper class which holds all user sessions and also provides
   method to get or create session.
 */
public class SessionStorage {

        // A map of user id and   sessions.
        HashMap<String, Session> userSession = new
        HashMap<String, Session>();

        /*
         * This method first checks if there is a session for
           this user (user ID). It also checks if the session is
           valid.
         *If there is no session for that user or if the session
           has expired, it will create a new session. Else it will
           return the active session.
         */
```

```java
public Session getOrCreateSession(String userId) {

    if (!userSession.containsKey(userId) ||
    !userSession.get(userId).isValid()) {
            Session session = new Session();
            userSession.put(userId, session);
    }
    return userSession.get(userId);
}

/**
 * Gets a String value from the session (if it exists) or
   the slot (if a match exists).
 *
 * @param match
 *          The intent match.
 * @param session
 *          The session.
 * @param slotName
 *          The name of the slot.
 * @param defaultValue
 *          The default value if not a value found in
            the session or slot.
 * @return The string value.
 */
public static String getStringFromSlotOrSession(Matched
Intent match, Session session, String slotName,
            String defaultValue) {
    String sessionValue = (String) session.
    getAttribute(slotName);
    if (sessionValue != null) {
            return sessionValue;
    }
```

```java
        return getStringSlot(match, slotName,
        defaultValue);
    }

    /**
     * Gets a String based slot value from an intent match.
     *
     * @param match
     *              The intent match to get the slot value from.
     * @param slotName
     *              The name of the slot.
     * @param defaultValue
     *              The default value to use if no slot found.
     * @return The string value.
     */
    public static String getStringSlot(MatchedIntent match,
    String slotName, String defaultValue) {
        if (match.getSlotMatch(slotName) != null && match.
        getSlotMatch(slotName).getMatchedValue() != null) {
                return (String) match.
                getSlotMatch(slotName).getMatchedValue();
        } else {
                return defaultValue;
        }
    }

    /**
     * Saves all the matched slots for an IntentMatch into
       the session.
     *
     * @param match
     *              The intent match.
     * @param session
```

```
*          The session.
*/
public static void saveSlotsToSession(MatchedIntent
match, Session session) {
        for (MatchedSlot matchedSlot : match.
        getSlotMatches().values()) {
                session.setAttribute(matchedSlot.getSlot().
                getName(), matchedSlot.getMatchedValue());
        }
    }
}
```

So far we showed how to create intent and slot classes and the matchers, and we discussed IRIS long-term and short-term memory. We will now discuss an essential concept for chatbot: conversation management. In the next section, we will explain how conversations can be modeled as finite state machines and used in IRIS.

Dialogues as Finite State Machines

Typically, a simple Q&A-based chatbot or a FAQ-based chatbot is not capable of having a conversation. A conversational chatbot should support complex dialog flow between the user and the bot, and we aim to build a chatbot that can mimic human conversation as much as possible. Usually, chatbots are limited to a request-response based flow and are not driven as dialog or conversations.

Building a chatbot by conversational state management helps in transitioning from one state to another. As shown in Figure 6-2, a state machine reads a series of inputs and switches to another state once it receives an input required to perform that transition.

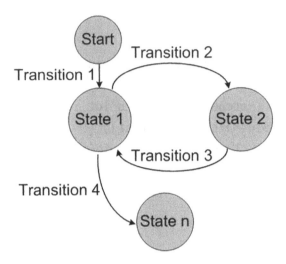

Figure 6-2. *Showing a finite state machine with states and transitions between states*

We'll explain this with a simple example. In Figure 6-3, we have a finite state representation of switching off and on a light bulb. There are two states: OFF and ON.

As in a finite state graph, you can only be in one state at a time. In the example of a light bulb, as shown in Figure 6-3, either it can be OFF or ON but not both at the same time. Also, to move from one state to another, a transition must take place. If the bulb is in the OFF state and we need to transition to ON state, we need to flip the switch up, which is an action/condition/prerequisite to transition to the ON state.

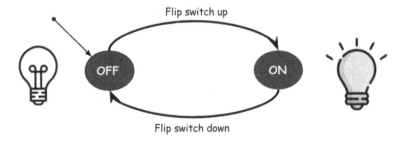

Figure 6-3. *State machine for a switch ON and OFF*

A state machine has the following components:

- **States**: Different states a bot can be in and transition to.

- **Initial State**: This is the start state when the user first interacts with IRIS.

- **Transitions**: The action(s) that should trigger a possible state change.

- **Shields**: A prerequisite or condition to transition to a target state.

A state machine can be designed in multiple ways. It can be modeled as a graph, the conversation could be modeled as a script, or it could be implemented using a very naive approach like HashMaps and some classes we will discuss next.

We need to create a new package in our project called com.iris.bot. state to contain the base classes for the state machine.

State

Let's first define State:

```
/*
 * State is an abstract class. Concrete State classes implement
   an execute method which is triggered when a transition to
   that state happens.
 */
public abstract class State {
    String name;

    public String toString() {
        return name;
    }
}
```

```java
    public State(String stateName) {
        name = stateName;
    }

    public String getName() {
        return name;
    }

    /*
     * The execute method takes a session and matched intent
       as an argument. The action of the state is defined in
       this method.
     */
    public abstract String execute(MatchedIntent matchedIntent,
    Session session);

}
```

The transition from one state to another may sometimes require a validation condition. We'll explain this with an example. If you are searching for a restaurant of your choice by interacting with a restaurant table booking chatbot, you can ask for cancellation only if you have booked a table at a restaurant. Otherwise, you cannot reach the state of cancellation.

Shields

To maintain any preconditions that may include business logic, we have Shield, which validates whether the transition to the desired state is possible or not.

```java
/*
 * Shield is an interface. The class that implements Shield
   implements the validate method and returns true if the
   validation condition is met.
```

```
 * Otherwise false will be returned and transition to that
   state will not happen.
 */
public interface Shield {

     public boolean validate(MatchedIntent match, Session
     session);

}
```

Transition

It is the Transition class that holds Shield and the target state information. The transition is an elementary class with two member variables, toState and shield, and their getters and setters as defined here:

```
/*
 * Transition class holds target state and shield information
 */
public class Transition {

     private State toState;
     private Shield shield;

/*
 * Constructor with target State (toState) and shield being set.
 */
     public Transition(State toState, Shield shield) {
          super();
          this.shield = shield;
          this.toState = toState;
     }

     public State getToState() {
          return toState;
     }
```

```java
    public void setToState(State toState) {
        this.toState = toState;
    }

    public Shield getShield() {
        return shield;
    }

    public void setShield(Shield shield) {
        this.shield = shield;
    }
}
```

State Machine

All of these pieces are stitched together in StateMachine. StateMachine is the backbone of state management in IRIS and knows the start state, the list of defined states, all the state transitions that are defined in the IRIS configuration, and has methods to add a state, add a transition, and most importantly, trigger the execution of the execute method upon successful transition:

```java
/*
 * StateMachine is the backbone class for IRIS state
   management. It contains start state, a map of states,
 * and a map of state transitions, all of which are defined in
   the Iris Configuration.
 */
public class StateMachine {
    /*
     * In start state, there will always be a predefined start
       state which will be the initial conversation state.
```

```
 * Start state is initialized in Iris configuration class.
 */
private State startState;

// A map of all the defined states.
private HashMap<String, State> states = new
HashMap<String, State>();

// A map of transition key and a list of possible transitions.
private HashMap<String, List<Transition>> stateTransitions =
new HashMap<String, List<Transition>>();

public void setStartState(State state) {
        this.startState = state;
}

// Method to add states in the state map.
private void addState(State state) {
        states.put(state.getName(), state);
        if (startState == null) {
                startState = state;
        }
}
/*
 * The addTransition method is used to add a transition
   from one state to another. It requires intent name,
   from state, and to state to define the transition.
 */
public void addTransition(String intentName, State
fromState, State toState) {

// When no Shield is passed, it is passed as null.
        addTransition(intentName, fromState, toState, null);
}
```

```
/*
 * Overloaded addTransition method that is similar to above
   but Shield is to be validated for this transition.
 */
public void addTransition(String intentName, State
fromState, State toState, Shield shield) {
        if (!states.containsKey(fromState.getName())) {
                addState(fromState);
        }

        if (!states.containsKey(toState.getName())) {
                addState(toState);
        }

        String key = makeTransitionKey(intentName, fromState);
        List<Transition> transitionList = stateTransitions.
        get(key);
        if (transitionList == null) {
                transitionList = new ArrayList<Transition>();
                stateTransitions.put(key, transitionList);
        }

        transitionList.add(new Transition(toState, shield));

}

/*
 * This method is the heart of the state machine. It
   receives the matched intent as an input along with
   session to know the current state. It then does a
   series of things: obtains the current state from a
   session or initializes the start state if no current
   state, then gets the matched intent, generates a
   transition key to look up in the transition map, and
```

> *finally triggers the execute method of the target*
> *state and updates the state in session.*
> */*

```
public String trigger(final MatchedIntent matchedIntent,
final Session session) {

        State currentState = startState;
```
// Gets the current state from the session. If it is a new
session, this will be null.
```
        String currentStateName = (String)
        session.getAttribute("currentStateName");

        if (currentStateName != null) {
                currentState = states.get(currentStateName);
```
// At this point, the current state should not be null, and
hence an exception is thrown as the handling of this condition
is unknown.
```
                if (currentState == null) {
                        throw new IllegalStateException("Illegal
                        current state in session:" +
                        currentStateName);
                }
        }

        Intent intent = matchedIntent.getIntent();
        String intentName = (intent != null) ? intent.
        getName() : null;
```
// intent should not null here as it is expected that the
matched intent will be an intent from the defined intent list.
```
        if (intentName == null) {
                throw new IllegalArgumentException("Request
                missing intent." + matchedIntent.toString());
        }
```

// Generate transition key by using the pattern "intentname-statename".

```
        String key = makeTransitionKey(intentName,
        currentState);
```

// Get the target state transition list from the state transitions map.

```
        List<Transition> transitionToStateList =
        stateTransitions.get(key);

        /*
         * If there is a condition where the intent is
           valid and the current state is valid but there
           is no transaction defined, and if there is no
           definition of where to go, it is an illegal
           state condition and cannot be handled.
         */
        if (transitionToStateList == null) {
                throw new IllegalStateException("Could not
                find state to transition to. Intent: " +
                intentName
                        + " Current State: " +
                        currentState);
        }

        State transitionToState = null;
```

// Find first matching to-state and check shield conditions. This method iterates one by one to find a successful transition target state.

```
            for (Transition transition : transitionToStateList) {
                if (transition.getShield() == null) {
```

```
// If there is no shield condition and there is a valid
transition, assign the transitionToState as that target state.
                    transitionToState = transition.
                    getToState();
                    break;
                } else {
// If there is a shield condition, it will be validated and
upon successful validation, the target state will be assigned
as transitionToState.
                    if (transition.getShield().
                    validate(matchedIntent, session)) {
                        transitionToState = transition.
                        getToState();
                        break;
                    }
                }
            }

// If state machine didn't find any matching states, it is an
illegal state as it is not defined.
        if (transitionToState == null) {
                throw new IllegalStateException("Could not
                find state to transition to. Failed all
                guards. Intent: "
                            + intentName + " Current State: "
                            + currentState);
        }

// Action to be performed upon successful transition and
response returned.
        String response = transitionToState.
        execute(matchedIntent, session);
```

```java
// Current state is now updated in the session.
        session.setAttribute("currentStateName",
        transitionToState.getName());

        return response;
    }

    /*
     * Transition key is defined to store transition key and
       a list of transitions.
     */
    private String makeTransitionKey(String intentName, State
    state) {
        return intentName + '-' + state.getName();
    }
}
```

We are done with defining the base classes and their implementations. Until now whatever we discussed formed the core of the IRIS framework. Now let's go further with a sample business use case and use the details to create specific intent classes, their slots, different states, and their possible transitions.

Building a Custom Chatbot for an Insurance Use Case

We discussed in Chapter 1 some of the most common applications of a chatbot in the life insurance industry. Now that we have some idea of the IRIS core, let's dive into building an insurance-focused chatbot using the IRIS framework.

At the end of this exercise, our chatbot should be capable of providing

- Account balance

- Life insurance quotation

- Claim status

- An advisor

- Answers to general enquiries

- Market trends

- Stock prices

- Weather details

The high-level functional architecture is described in Figure 6-4. There are communication client channels such as Facebook Messenger, web chat, and Alexa, via which the users can connect to IRIS. In Figure 6-4, a *channel integration module* acts as a gateway module. It integrates with services such as Facebook Messenger, receives the request, and delegates the request for IRIS to respond. The response is sent back to Messenger by this module.

Then there is the IRIS Engine Core that handles all the domain-specific business logic that controls the behavior of the chatbot platform as well as defines and manages the transition from one state to another. Core connects with the *intent classification engine* that predicts the intent from a user's utterance. The capabilities of IRIS that we discussed require an information retrieval module that can query its semantic knowledge base, a quotation service that provides life insurance quotations based on user inputs, a website search service, a user module that connects to a user database to get account balance information, a claims module to fetch claim details, and other third-party services to get market trends, stock prices, and weather information.

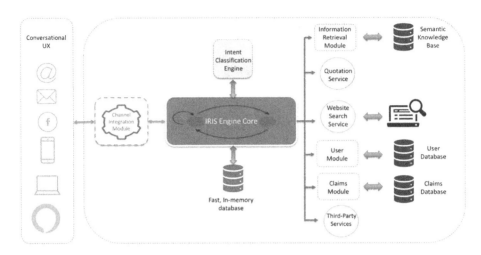

Figure 6-4. *High-level function architecture*

We require the following intents to be defined for our example use case:

- AccountBalanceIntent

- AskForQuoteIntent

- ClaimStatusIntent

- ExitIntent

- FindAdvisorIntent

- GeneralQueryIntent

- GetAccTypeIntent

- GetClaimIdIntent

- MarketTrendIntent

- StockPriceIntent

- WeatherIntent

Creating the Intents

Let's create a User Intent class named AccountBalanceIntent:

```
public class AccountBalanceIntent extends Intent {
    public AccountBalanceIntent() {
        super("accountBalanceIntent");
}}
```

All the other intent classes are created in the same way with their intent names. Some of these intents will have one or more slots defined as well.

AskForQuoteIntent requires four slots to provide a life insurance quote in our example:

- Age (CustomNumericSlot type)

- Height (CustomNumericSlot type)

- Smoker (BooleanLiteralSlot type)

- Weight (CustomNumericSlot type)

AccountBalanceIntent requires two slots: account type for which the account balance is required and a user PIN to authenticate the user. ipin is just a way of demonstrating how a very basic authentication can be performed. In an actual implementation, more complex forms of authentication should be used.

- Account Type (AccTypeSlot type)

- ipin (IPinSlot type)

ClaimStatusIntent and GetClaimIdIntent intents require claimId (AlphaNumericSlot type)

CustomNumericSlot

Next, let's see the code showing how the above mentioned slot types are implemented for providing a simple approach to fulfill our requirements:

```
/*
 * Custom numeric slot
 */
public class CustomNumericSlot extends Slot {
    private String name;

    public CustomNumericSlot(String name) {
        super();
        this.name = name;
    }

    /*
     * match CustomNumericSlot takes user utterance and
       returns MatchedSlot if there is a slot match.
     * In this method we use regex and hard-coded words to
       number logic to identify if there is a number.
     * example - 18, eighteen
     */
    public MatchedSlot match(String utterance) {
        String token = utterance.replaceAll("[^0-9]+", "");
        if (token.isEmpty()) {
            token = String.valueOf(wordStringToNumber
            (utterance));
        }
        return new MatchedSlot(this, token, token.
        toLowerCase());
    }
```

```java
/*
 * This method converts words to numbers. The logic is derived
   from https://stackoverflow.com/questions/26948858/converting-
   words-to-numbers-in-java.
*/
        public Number wordStringToNumber(String wordString) {
            if (wordString == null || wordString.length() < 1) {
                return null;
            }

            wordString = wordString.replaceAll("-", " ");
            wordString = wordString.replaceAll(",", " ");
            wordString = wordString.toLowerCase().replaceAll
            (" and", " ");
            String[] splittedParts = wordString.trim().
            split("\\s+");

            long finalResult = 0;
            long result = 0;

            for (String str : splittedParts) {
                if (str.equalsIgnoreCase("zero")) {
                    result += 0;
                } else if (str.equalsIgnoreCase("one")) {
                    result += 1;
                } else if (str.equalsIgnoreCase("two")) {
                    result += 2;
                } else if (str.equalsIgnoreCase("three")) {
                    result += 3;
                } else if (str.equalsIgnoreCase("four")) {
                    result += 4;
```

```java
    } else if (str.equalsIgnoreCase("five")) {
        result += 5;
    } else if (str.equalsIgnoreCase("six")) {
        result += 6;
    } else if (str.equalsIgnoreCase("seven")) {
        result += 7;
    } else if (str.equalsIgnoreCase("eight")) {
        result += 8;
    } else if (str.equalsIgnoreCase("nine")) {
        result += 9;
    } else if (str.equalsIgnoreCase("ten")) {
        result += 10;
    } else if (str.equalsIgnoreCase("eleven")) {
        result += 11;
    } else if (str.equalsIgnoreCase("twelve")) {
        result += 12;
    } else if (str.equalsIgnoreCase("thirteen")) {
        result += 13;
    } else if (str.equalsIgnoreCase("fourteen")) {
        result += 14;
    } else if (str.equalsIgnoreCase("fifteen")) {
        result += 15;
    } else if (str.equalsIgnoreCase("sixteen")) {
        result += 16;
    } else if (str.equalsIgnoreCase("seventeen")) {
        result += 17;
    } else if (str.equalsIgnoreCase("eighteen")) {
        result += 18;
    } else if (str.equalsIgnoreCase("nineteen")) {
        result += 19;
    } else if (str.equalsIgnoreCase("twenty")) {
```

```
        result += 20;
    } else if (str.equalsIgnoreCase("thirty")) {
        result += 30;
    } else if (str.equalsIgnoreCase("forty")) {
        result += 40;
    } else if (str.equalsIgnoreCase("fifty")) {
        result += 50;
    } else if (str.equalsIgnoreCase("sixty")) {
        result += 60;
    } else if (str.equalsIgnoreCase("seventy")) {
        result += 70;
    } else if (str.equalsIgnoreCase("eighty")) {
        result += 80;
    } else if (str.equalsIgnoreCase("ninety")) {
        result += 90;
    } else if (str.equalsIgnoreCase("hundred")) {
        result *= 100;
    } else if (str.equalsIgnoreCase("thousand")) {
        result *= 1000;
        finalResult += result;
        result = 0;
    } else if (str.equalsIgnoreCase("million")) {
        result *= 1000000;
        finalResult += result;
        result = 0;
    } else if (str.equalsIgnoreCase("billion")) {
        result *= 1000000000;
        finalResult += result;
        result = 0;
    } else if (str.equalsIgnoreCase("trillion")) {
        result *= 1000000000000L;
```

```
                    finalResult += result;
                    result = 0;
            } else {
                    // unknown word
                    return null;
            }
        }

        finalResult += result;
        result = 0;
        return finalResult;
    }

    @Override
    public String getName() {
        return name;
    }
}
}
```

BooleanLiteralSlot

Here we've highlighted a match method snippet of the
BooleanLiteralSlot class:

```
/*
* match method of BooleanLiteralSlot. We need to recognize
  if the user meant no or yes in any which way. One of the
  simplest ways to implement this is to verify by string
  matching the most commonly used words.
*/
    @Override
    public MatchedSlot match(String utterance) {
```

```java
        if (utterance.toLowerCase().contains("yes") ||
        utterance.toLowerCase().contains("yeah")
                    || utterance.toLowerCase().
                    contains("ya") || utterance.
                    toLowerCase().contains("yup")) {
            return new MatchedSlot(this, utterance, "yes");
        } else if (utterance.toLowerCase().contains("no")
        || utterance.toLowerCase().contains("na")
                    || utterance.toLowerCase().
                    contains("nopes") || utterance.
                    toLowerCase().contains("noo")
                    || utterance.toLowerCase().
                    contains("nope") || utterance.
                    toLowerCase().contains("dont")
                    || utterance.toLowerCase().
                    contains("don't") || utterance.
                    toLowerCase().contains("do not")) {
            return new MatchedSlot(this, utterance, "no");
        }
        return null;
    }
```

AccTypeSlot

AccTypeSlot is implemented to understand the account type. If no slot
match happens, the state engine will reprompt as the system could not
identify the account type:

```java
    /*
     * For intent where we want to understand what type
       of account balance the user is looking for, a
       straightforward method is to apply a
```

```
     * string match of possible account types. Since we are
       only looking for whether the utterance contains any of
       those keywords,
     * all of the below possibilities are covered:
     * i am looking for annuities account balance
     * annuities
     * annuities balance
     * tell 401k balance
     * want my retirement balance etc.
     */
    @Override
    public MatchedSlot match(String utterance) {
            if (utterance.toLowerCase().contains("annuities")
            || utterance.toLowerCase().contains("annuity")) {
                    return new MatchedSlot(this, "annuities",
                    "annuities");
            } else if (utterance.toLowerCase().contains("401k")
            || utterance.toLowerCase().contains("retirement")
                         || utterance.toLowerCase().
                         contains("401") || utterance.
                         toLowerCase().contains("401 k")) {
                    return new MatchedSlot(this, "401k", "401k");
            }
            return null;
    }
```

IPinSlot

In an actual implementation, this type of slot may not be defined but to highlight how a basic authentication can be implemented, we use this entity. We have considered in the example that users will have their own ipin generated in some way and stored in the back end and that it will be a six- digit number. In a real world, a much more complex number and a set of authentication mechanisms will exist such as username, password, and ZIP code.

In the method snippet below, if the value is 123456, only then will the account balance be displayed. Any other number will result in a wrong ipin provided by the user.

Warning   Never implement such a weak authentication system. It will compromise your enterprise security. The purpose here is to only complete the flow of discussion. In no way do we endorse such weak authentication.

```java
@Override
    public MatchedSlot match(String token) {
        if (token.matches("[0-9]+") && token.length() == 6
        && token.equalsIgnoreCase("123456")) {
            return new MatchedSlot(this, token, token);
        }
        return null;
    }
```

AlphaNumericSlot

As the name suggests, the entity is supposed to be alphanumeric, and if there is a word that is alphanumeric, MatchedSlot is returned. In the example, AphaNumericSlot is used for processing claims that are alphanumeric.

```java
@Override
    public MatchedSlot match(String utterance) {
        /*
```

```
* User utterance is split into utterance tokens.
  We need to see if there is any alphanumeric word
  in the utterance
* This implementation is useful for scenarios
  mentioned below:
* my claim id is gi123 can you tell the claim status
* claim the status of abc123
*/
ArrayList<String> utteranceTokens = new
ArrayList<String>(Arrays.asList(utterance.split
("\\s+")));
String claimId = null;
for (String token : utteranceTokens) {
        if (!token.matches("[a-zA-Z]+")) {
                token = token.replace(".", "");
                token = token.trim();
                claimId = token;
                return new MatchedSlot(this, claimId,
                claimId);
        }
    }
    return null;
}
```

Now that we have defined all the possible intents that will be classified by our intent classification service, defined slots, and slot type, let's see what the IRIS configuration looks like. As explained, intents, intent matcher, slot matcher, and different slot and slot types are defined in the IrisConfiguration class.

IrisConfiguration

We put IrisConfiguration class in a separate package named com.iris. bot.config:

```java
public class IrisConfiguration {

    public IntentMatcherService getIntentMatcherService() {
        CustomSlotMatcher slotMatcher = new
        CustomSlotMatcher();
        IntentMatcherService intentMatcherService = new Int
        entMatcherService(slotMatcher);

        Intent findAdvisorIntent = new FindAdvisorIntent();

        Intent askForQuoteIntent = new AskForQuoteIntent();
        // Slots for askForQuote intent fulfillment.
        askForQuoteIntent.addSlot(new
        CustomNumericSlot("age"));
        askForQuoteIntent.addSlot(new
        CustomNumericSlot("height"));
        askForQuoteIntent.addSlot(new
        CustomNumericSlot("weight"));
        askForQuoteIntent.addSlot(new
        BooleanLiteralSlot("smoked"));

        Intent generalQueryIntent = new GeneralQueryIntent();

        Intent stockPriceIntent = new StockPriceIntent();

        Intent marketTrendIntent = new MarketTrendIntent();

        Intent accountBalanceIntent = new
        AccountBalanceIntent();
        accountBalanceIntent.addSlot(new
        AccTypeSlot("accType"));
```

243

```
            accountBalanceIntent.addSlot(new IpinSlot("ipin"));

            Intent getAccTypeIntent = new GetAccTypeIntent();
            getAccTypeIntent.addSlot(new
            AccTypeSlot("accType"));
            getAccTypeIntent.addSlot(new IpinSlot("ipin"));

            Intent weatherIntent = new WeatherIntent();

            Intent claimStatusIntent = new ClaimStatusIntent();
            claimStatusIntent.addSlot(new
            AlphaNumericSlot("claimId"));

            Intent getClaimIdIntent = new GetClaimIdIntent();
            getClaimIdIntent.addSlot(new
            AlphaNumericSlot("claimId"));

            Intent exitIntent = new ExitIntent();
/*
 * All the intents we defined above are added to the intent
   matcher service.
 */
            intentMatcherService.addIntent(findAdvisorIntent);
            intentMatcherService.addIntent(askForQuoteIntent);
            intentMatcherService.addIntent(generalQueryIntent);
            intentMatcherService.addIntent(stockPriceIntent);
            intentMatcherService.addIntent(marketTrendIntent);
            intentMatcherService.addIntent(exitIntent);
            intentMatcherService.addIntent(getAccTypeIntent);
            intentMatcherService.addIntent(accountBalanceIntent);
            intentMatcherService.addIntent(weatherIntent);
            intentMatcherService.addIntent(claimStatusIntent);
            return intentMatcherService;
    }
```

```java
    public StateMachine getStateMachine() {
    // discussed in detail below
     return null;
  }
}
```

Adding States

Before we add state machine configurations, let's see how many possible states we have:

1. Start state

2. Ask for quote state

3. Get quote state   .

4. Find an advisor state

5. General query state

6. Stock price state

7. Market trend state

8. Get account balance state

9. Get account type state

10. Get weather state

11. Get claim status state

12. Exit state

The **getStateMachine** method is now added to IrisConfiguration with the states listed above:

```
public StateMachine getStateMachine() {
/*
Creates an instance of StateMachine that holds start state, a
map of states, and a map of state transitions, all of which are
defined below in IrisConfiguration.
 */
            StateMachine stateMachine = new StateMachine();
            State startState = new StartState();

            State askforQuoteState = new AskForQuoteState();
            State getQuoteState = new GetQuoteState();
            Shield haveQuoteDetailShield = new
            HaveQuoteDetailShield();
            Shield dontHaveQuoteDetailsShield = new
            DontHaveQuoteDetailsShield();

            State findAdvisorState = new FindAdvisorState();
            State generalQueryState = new GeneralQueryState();
            State stockPriceState = new StockPriceState();
            State marketTrendState = new MarketTrendState();

            State getAccountBalanceState = new
            GetAccountBalanceState();
            Shield haveAccTypeShield = new HaveAccTypeShield();
            Shield dontHaveAccTypeShield = new
            DontHaveAccTypeShield();
            State getAccTypeState = new GetAccTypeState();

            State getWeatherState = new GetWeatherState();

            State getClaimStatusState = new GetClaimStatus();
            Shield haveClaimIdShield = new HaveClaimIdShield();
```

```
        State getClaimIdState = new GetClaimIdState();

        State exitState = new ExitState();
/*
* Here we initialize the start state. The Start state execute
  method is never supposed to be called.
 */
        stateMachine.setStartState(startState);
/*
* We need to define state transitions here.
 */
    }
```

In **getStateMachine** method, we define the state classes and Shields.

Shields

As discussed, Shields provide a Boolean condition for transition from one state to another. If all the information required for transitioning to another state is available, Shields returns true.

We implement five shields in our example in the **getStateMachine** method, each implementing the **validate** method.

DontHaveAccTypeShield

We need accType and a valid ipin in order to transition to GetAccountBalanceState. This shield returns true if either of them is not provided.

```
public boolean validate(MatchedIntent match, Session session) {
        // save slots to session
        SessionStorage.saveSlotsToSession(match, session);
```

```
// Get all validation entities from session.
String accType = SessionStorage.getStringFromSlotOrSession(match,
session, "accType", null);
            String ipin = SessionStorage.getStringFromSlotOr
            Session(match, session, "ipin", null);

// Returns true if accType or ipin is null.
            return (accType == null || ipin == null);
        }
```

DontHaveQuoteDetailsShield

We need age, smoker info, height, and weight to provide insurance quotation eligibility. This shield returns true if we do not have information on any of them. The state remains in AskForQuoteState until we have answers for all questions and then transitions to GetQuoteState.

```
Public boolean  validate(MatchedIntent match, Session session) {

// Save slots into session.
            SessionStorage.saveSlotsToSession(match, session);

            String age = SessionStorage.getStringFromSlotOr
            Session(match, session, "age", null);
            String smoked = SessionStorage.getStringFromSlotOr
            Session(match, session, "smoked", null);
            String height = SessionStorage.getStringFromSlotOr
            Session(match, session, "height", null);
            String weight = SessionStorage.getStringFromSlotOr
            Session(match, session, "weight", null);

/*
 * If we don't have all the slots fulfilled, we need to return
   true so that askForQuote state is executed again. As there are
   multiple questions asked in the askForQuote state, unless all
```

questions are answered and all values populated, the state
remains the same, unless the intent of the user changes.
*/

```
            return (age == null || smoked == null || height ==
            null || weight == null);
    }
```

HaveAccTypeShield

This shield returns true if both accType and ipin are available from the
user and allows a transition to the GetAccountBalanceState:

```
Public boolean validate(MatchedIntent match, Session session) {
            SessionStorage.saveSlotsToSession(match, session);
            String accType = SessionStorage.getStringFromSlotOr
            Session(match, session, "accType", null);
            String ipin = SessionStorage.getStringFromSlotOr
            Session(match, session, "ipin", null);
// Returns true only if both accType and ipin are available.
            return (accType != null && ipin != null);
    }
```

HaveClaimIdShield

This shield returns true if claimId is not null and hence allows a transition
to the GetClaimStatus state:

```
public boolean validate(MatchedIntent request, Session session) {
            SessionStorage.saveSlotsToSession(request, session);
            String claimId = SessionStorage.getStringFromSlotOr
            Session(request, session, "claimId", null);
// Returns true only if claimId is not null.
            return (claimId != null);
    }
```

HaveQuoteDetailShield

This shield returns true if all the values required to transition to
GetQuoteState are present:

```java
public boolean validate(MatchedIntent match, Session session) {
        // Saves slots to session.
        SessionStorage.saveSlotsToSession(match, session);

        // Gets all validation entities from session.
        String age = SessionStorage.getStringFromSlotOr
        Session(match, session, "age", null);
        String smoked = SessionStorage.getStringFromSlotOr
        Session(match, session, "smoked", null);
        String height = SessionStorage.getStringFromSlotOr
        Session(match, session, "height", null);
        String weight = SessionStorage.getStringFromSlotOr
        Session(match, session, "weight", null);

        //Returns true if all values exist, else return false.
        return (age != null && smoked != null && height !=
        null && weight != null);
    }
```

We have two more things yet to be discussed to complete
IrisConfiguration related concepts:

- The execute method of each state

- State transitions

Adding Execute Methods

Let's start by implementing the execute method of each state described in
the example.

Exit State

The execute method of ExitState responds with a simple reply. In an actual implementation, it could also support saving the session and context to a persistent storage before resetting them.

```java
public class ExitState extends State {
    public ExitState() {
        super("exitState");
    }
/*
 * When this state is reached, the execute method is invoked.
   As a result, a reply is sent back.
 */

    @Override
    public String execute(MatchedIntent matchedIntent,
    Session session) {
        String reply = "Anything else that I may help you
        with?";
        return reply;
    }
}
```

FindAdvisorState

The execute method of FindAdvisorState would typically call a search API with required parameters such as advisor name, ZIP code, etc. to return relevant advisors to the user. We demonstrate how to reach here but we skip the implementation.

```java
public String execute(MatchedIntent matchedIntent, Session session) {
        String reply = "You know what, I dont have the data
        about financial advisors with me."
```

```
                    + "\nBut I hope you do get the point
                    that I could have surely provided
                    it to you if I was connected to a
                    database.\n"
                    + "I will let my boss know that you
                    were asking for it. Next time you wont
                    be disappointed, I promise.\n Here,
                    ask me anything else for now please!";
        return reply;
    }
```

GetAccountBalanceState

The execute method of this state returns the account balance. Transition to this state only happens when the shield validates that we have the account type and ipin.

```
    Public String execute(MatchedIntent matchedIntent,
    Session session) {
            String reply = null;
            Random rand = new Random();
            String accType = SessionStorage.getStringFromSlotOr
            Session(matchedIntent, session, "accType", null);
            if (accType.equalsIgnoreCase("Annuities")) {
/*
* In a real-world implementation, we would call a service or
  query a database to get the account balance. For the sake of
  implementation here, we are returning a random integer.
*/
                    reply = "Your Annuities account balance is: "
                    + (rand.nextInt(1000) + 100) + "."
                            + "\nAnything else that I can do
                            for you? ";
```

```
      } else if (accType.equalsIgnoreCase("401k"))
            reply = "Your 401K account balance is: " +
            (rand.nextInt(4000) + 500) + "."
                        + "\nAnything else that you want
                        to know? ";
      else
            reply = "Sorry, I am not able to retrieve
            your " + accType + " balance right now.\nHow
            else can I help you? ";
```

```
/*
 * Slot details saved in session attributes previously are now
   removed. We cannot store these details even at a session
   level as the user may request for account balance again, but
   this time he may need balance details of a different type
   of account. However, we still store these values in session
   until we reach here so that we know that this information
   have been answered by user and shields can then validate.
*/
            session.removeAttribute("acctype");
            session.removeAttribute("getaccTypeprompt");
            session.removeAttribute("getipinprompt");
            session.removeAttribute("ipin"s);

            return reply;
      }
```

GetAccTypeState

The execute method of this state prompts for the ipin and account type information to be provided by user.

```java
public String execute(MatchedIntent matchedIntent,
Session session) {
        SessionStorage.saveSlotsToSession(matchedIntent,
        session);

        String reply = null;

        if (SessionStorage.getStringFromSlotOrSession
        (matchedIntent, session, "ipin", null) == null) {
                if (SessionStorage.getStringFromSlotOrSession
                (matchedIntent, session, "getipinprompt",
                null) == null)
                        reply = "Sure I will help you with
                        that! Since this is a confidential
                        information, I will need additional
                        details to verify "
                                    + "your identity. Can you
                                    tell me your 6 digits IPIN
                                    please?";
                else
                    reply = "Either you have not entered 6
                    digits code or the IPIN entered by you is
                    incorrect. Please verify and type again !";
                session.setAttribute("getipinprompt", "flag1");
        }

        else if (SessionStorage.getStringFromSlotOrSession(
        matchedIntent, session, "accType", null) == null) {
                if (SessionStorage.getStringFromSlotOr
                Session(matchedIntent, session,
                "getaccTypeprompt", null) == null)
```

```
                          reply = "Your IPIN was successfully
                          verified. Are you looking for Annuities
                          balance or 401k account balance?";
                else
                          reply = "I did not understand that.
                          Did you say annuities or 401k?";
                session.setAttribute("getaccTypeprompt",
                "flag1");
        }
        return reply;
}
```

GetClaimIdState

The execute method of this state obtains a claim ID from the user's utterance and sets it in the session attribute.

```
public String execute(MatchedIntent matchedIntent,
Session session) {
        SessionStorage.saveSlotsToSession(matchedIntent,
        session);

        if (SessionStorage.getStringFromSlotOrSession
        (matchedIntent, session, "claimId", null) == null) {
                if (SessionStorage.getStringFromSlotOr
                Session(matchedIntent, session,
                "getclaimidprompt", null) == null)
                        reply = "No Problem. Could you tell me
                        the Claim Id Please?";
                else
                        reply = "Sorry, I did not get the claim
                        ID. Can you please re-enter it?";
        }
```

```
            session.setAttribute("getclaimidprompt", "flag1");
            return reply;
    }
```

AskForQuote State

The execute method of this state gets age, smoker info, height, and weight from user. It also stores the last question asked to map it back to the follow up answer.

```
    public String execute(MatchedIntent matchedIntent,
    Session session) {

            SessionStorage.saveSlotsToSession(matchedIntent,
            session);
// Default reply
            String reply = "I am having trouble understanding...";

// Checking for age
            if (SessionStorage.getStringFromSlotOrSession
            (matchedIntent, session, "age", null) == null) {
// Age is set in session to be the last question asked in
askQuote at this point.
                    session.setAttribute("askQuoteLastQuestion",
                    "age");
/*
* To differentiate between whether we are asking this question
  for the first time or we asked before and the user didn't
  answer, we use "getageprompt." If the "getageprompt" value
  is null, we have not asked this question to the user before
  in that particular session. It helps to differentiate the
  reply message.
*/
```

```
if (SessionStorage.getStringFromSlotOrSession(matchedIntent,
session, "getageprompt", null) == null)
                        reply = "Sure, I will help you with
                        that. May I know your age?";
            else
```

/*
* Let's say we are expecting that the user will enter his age
  and that is the current question in conversation. However,
  instead of replying age, the user changes the intent by
  asking about the weather. IRIS is designed to handle intent
  switches from one context to another. But, next time, if the
  user desires to get a quote again, we will not ask questions
  already answered and even the ask message will be different,
  just like how its mentioned in the if-else reply message here.
  */

```
                        reply = "I am not sure if I got your age
                        right last time. Please type again";
```

// Setting getageprompt in session to note that age has been
asked before.

```
                        session.setAttribute("getageprompt", "flag1");
```

// Same logic applies for whether the user answered to his
smoking status or not.

```
            } else if (SessionStorage.getStringFromSlotOrSession
            (matchedIntent, session, "smoked", null) == null) {
                    session.setAttribute("askQuoteLastQuestion",
                    "smoked");
                    if (SessionStorage.getStringFromSlotOrSession
                    (matchedIntent, session, "getsmokedprompt",
                    null) == null)
```

```
                        reply = "Have you smoked in the last
                        12 months?";
            else
                        reply = "Last time you did not tell me
                        if you smoked in the last 12 months,
                        Have you?";
                session.setAttribute("getsmokedprompt",
                "flag1");
```

```
// Same logic applies for height.
            } else if (SessionStorage.getStringFromSlotOrSession
            (matchedIntent, session, "height", null) == null) {
                session.setAttribute("askQuoteLastQuestion",
                "height");
                if (SessionStorage.getStringFromSlotOrSession
                (matchedIntent, session, "getheightprompt",
                null) == null)
                        reply = "What's your height (in
                        centimeters)?";
                else
                        reply = "What's your height (in
                        centimeters)? Please help me
                        understand again?";
                session.setAttribute("getheightprompt",
                "flag1");
```

```
// Lastly, same logic for weight.
            } else if (SessionStorage.getStringFromSlotOrSession
            (matchedIntent, session, "weight", null) == null) {
                session.setAttribute("askQuoteLastQuestion",
                "weight");
```

```
        if (SessionStorage.getStringFromSlotOrSession
        (matchedIntent, session, "getweightprompt",
        null) == null)
              reply = "What's your weight (in
              pounds)?";
        else
              reply = "Tell me your weight in pounds
              again. I did not get it the last time";
        session.setAttribute("getweightprompt",
        "flag1");
    }
    return reply;
}
```

GetQuote State

The execute method of this state provides quotation eligibility based on the age, smoker info, height, and weight. The method implements a simple business logic to calculate if the user is eligible or not. However, in a real-world scenario, more complex business logic exists and all of this information will be passed to another API that will provide the eligibility information.

```
Public String execute(MatchedIntent matchedIntent, Session
session) {
        SessionStorage.saveSlotsToSession(matchedIntent,
        session);

        Boolean eligible = true;
        String answer = "";

        int age = Integer.parseInt(SessionStorage.getString
        FromSlotOrSession(matchedIntent, session, "age",
        null));
```

```java
String smoked = SessionStorage.getStringFromSlotOr
Session(matchedIntent, session, "smoked", null);
int weight = Integer.parseInt(SessionStorage.get
StringFromSlotOrSession(matchedIntent, session,
"weight", null));
int height = Integer.parseInt(SessionStorage.get
StringFromSlotOrSession(matchedIntent, session,
"height", null));

/*
 * Checking business logic and calculating BMI
   (body mass index).
 * In the example, eligibility is defined based on
   whether BMI is less than or greater than 33.
 */
if (age > 60 || age < 18)
     eligible = false;
if (smoked.equalsIgnoreCase("yes"))
     eligible = false;

double weightInKilos = weight * 0.453592;
double heightInMeters = ((double) height) / 100;
double bmi = weightInKilos / Math.
pow(heightInMeters, 2.0);

if (bmi > 33)
     eligible = false;

if (eligible) {
     answer = "Great News! You are eligible for
     an accelerated UW Decision.\nPlease proceed
     with your application "
              + "at this link: https://www.
              dummylink.com \n"
```

```
                              + "Anything else that I could
                              help you with?";
        } else {
                answer = "Unfortunately, You are not eligible
                for an Accelerated UW Decision.\nPlease
                register at https://www.dummylink.com "
                              + "and our representatives
                              will contact with you shortly
                              to further process your
                              application\n"
                              + "Anything else that I could
                              help you with?";

        }
/*
* Remove attributes stored in the session. All of these four
  attributes are treated as short term in the example. We used
  the session also to store details of which slots to prompt
  for and which not to based on whether user answered them.
*/
        session.removeAttribute("getageprompt");
        session.removeAttribute("getsmokedprompt");
        session.removeAttribute("getheightprompt");
        session.removeAttribute("getweightprompt");
        session.removeAttribute("askquotelastquestion");

        session.removeAttribute("height");
        session.removeAttribute("age");
        session.removeAttribute("smoked");
        session.removeAttribute("weight");
        return answer;

    }
```

Start State

This state is the starting state and the "current" state by default when the user interacts in a new session. A start state is never executed due to the result of any behavior.

```
public String execute(MatchedIntent matchedIntent, Session
session) {
            throw new IllegalStateException("You shouldn't be
            executing this state!");
    }
```

GeneralQuery State

We mentioned that in a chatbot where there are multiple intents such as a user looking for account balance, claim status, weather details, life insurance quote, etc., the general query is not an explicit intent. We classify an utterance into a general query if no other intent matches explicitly.

In the general query state, we perform two steps:

1. Match if a user utterance is a question that has an answer in our knowledge repository. The knowledge repository is where the most frequently asked questions and their answers are stored. A knowledge repository could also have general user information parsed and stored in a way that can be queried to find a meaningful answer. A knowledge repository could be represented in the form of a graph, RDF semantic web, or implemented using a simple search engine.

2. If there is no matching answer in our knowledge repository, we perform a search on our portal to find any matching result that could be replied to the user. If there is no response from the search service as well,

we reply to the user saying we cannot help on this ask
because we don't have much information about it now.

```java
public String execute(MatchedIntent matchedIntent, Session
session) {
            String answer = "I am so Sorry, I do not have any
            information related to your query. Can I help you
            with something else?";
            String uri = "https://www.dummy-knowledge-base-
            service-url?inputString=";
            uri = uri + matchedIntent.getUtterance();

            RestTemplate restTemplate = new RestTemplate();
            String result = restTemplate.getForObject(uri,
            String.class);

            ObjectMapper mapper = new ObjectMapper();
            try {

                if(result!=null){
/*
* If the result is not null and contains a response (answer),
  we parse that information and assign it to the answer
  variable answer = "PARSED-INFORMATION from result" + "\
  nAnything else that you would like to ask?";
 */
}
                else{
/*
* If no answer was obtained from the knowledge repository, then
  to back fill with some valid response, we call the enterprise
  search API and pass the utterance as a search string.
*/
```

```
                        uri = "https://www.my-enterprise-
                        website.com/searchservice/fullsearch?&
                        inputSearchString=";
/*
 * The utterance is added to the HTTP GET request. Depending on
   the implementation it could be GET or POST and the service
   may have different parameters.
 */
                        uri = uri + matchedIntent.
                        getUtterance();
                        result = restTemplate.
                        getForObject(uri, String.class);
                        mapper = new ObjectMapper();
                        try {
/*
 * Here we try to parse the JSON response and if there is a
   result with a decent score returned from the search engine,
   we read the title and description of the result and add it
   before sending the response back.
 */
                        answer = "Sorry I do not have an
                        exact answer to this right now. "
                                    + "You may get
                                    some details on
                                    the page - " +
                                    "TITLE OF THE PAGE
                                    OBTAINED FROM
                                    RESPONSE"
                                    + ". Click here ->
                                    " + "URL LINK" + "
                                    for more info."
```

```
                                   + "\nAnything else
                                   that you would like
                                   to ask?";

            } catch (Exception e) {
                    answer = "I am so Sorry, I do
                    not have any information related
                    to your query. Can I help you
                    with something else?";

            }

        }
    } catch (Exception e) {
            e.printStackTrace();
            answer = "I am so Sorry, I do not have any
            information related to your query. Can I
            help you with something else?";
    }
    return answer;
}
```

Market trends, stock prices, weather state, and claim status state require integration with third-party data sources or connecting to a database. We will discuss this in the next chapter in detail.

Adding State Transitions

In the getStateMachine method of the IrisConfiguration class, we define the transitions from one state to another. For example, we can transition to any state from a start state, as explained in the following snippet. The first argument of the addTransition method is the intent name, the second argument is the current state, the third argument is the target state, and the fourth argument is an optional shield.

In this example, since we are defining transitions from startState and fromState, all transitions will be startState.

```
/*
 * This transition says that if we are in the start
   state, and a generalQueryIntent is obtained,
 * we remain in the generalQueryState (and trigger
   execute method of this state).
 */
stateMachine.addTransition("generalQueryIntent",
startState, generalQueryState);
/*
 * This transition says that if we are in the
   start state, and an askForQuoteIntent intent is
   obtained,
 * we change to the target state which is
   getQuoteState if the shield conditions are
   validated.
 * Else we check the next transition condition.
 */
stateMachine.addTransition("askForQuoteIntent",
startState, getQuoteState, haveQuoteDetailShield);
/*
 * If the shield conditions are not validated for
   the askForQuoteIntent, it means that we do not
   have all the information
 * for switching to getQuoteState, which provides
   quote details. Hence, in that case, we switch to
   askforQuoteState without the need
 * of a shield.
 */
```

```
stateMachine.addTransition("askForQuoteIntent",
startState, askforQuoteState);
stateMachine.addTransition("findAdvisorIntent",
startState, findAdvisorState);
stateMachine.addTransition("stockPriceIntent",
startState, stockPriceState);
stateMachine.addTransition("marketTrendIntent",
startState, marketTrendState);

/*
 * If we are in the start state and the user
   intends to get an account balance, we validate
   with a shield if we have an account type
 * and ipin details to switch to getAccountBalanceState
   and trigger its execute method.
 */
stateMachine.addTransition("accountBalanceIntent",
startState, getAccountBalanceState,
haveAccTypeShield);
/*
 * Otherwise, if shield does not validate, it means
   we do not have all the details and hence we
   switch to getAccTypeState
 * to get all the details.
 */
stateMachine.addTransition("accountBalanceIntent",
startState, getAccTypeState);
stateMachine.addTransition("weatherIntent",
startState, getWeatherState);
stateMachine.addTransition("claimStatusIntent",
startState, getClaimStatusState,
haveClaimIdShield);
```

```
stateMachine.addTransition("claimStatusIntent",
startState, getClaimIdState);
```

Similarly, we can define state transitions from the findAdvisor state:

```
stateMachine.addTransition("exitIntent",
findAdvisorState, exitState);
stateMachine.addTransition("marketTrendIntent",
findAdvisorState, marketTrendState);
stateMachine.addTransition("findAdvisorIntent",
findAdvisorState, findAdvisorState);
stateMachine.addTransition("askForQuoteIntent",
findAdvisorState, askforQuoteState);
stateMachine.addTransition("generalQueryIntent",
findAdvisorState, generalQueryState);
stateMachine.addTransition("weatherIntent",
findAdvisorState, getWeatherState);
stateMachine.addTransition("claimStatusIntent",
findAdvisorState, getClaimStatusState,
haveClaimIdShield);
stateMachine.addTransition("claimStatusIntent",
findAdvisorState, getClaimIdState);
stateMachine.addTransition("accountBalanceIntent",
findAdvisorState, getAccountBalanceState,
haveAccTypeShield);
stateMachine.addTransition("accountBalanceIntent",
findAdvisorState, getAccTypeState);
stateMachine.addTransition("stockPriceIntent",
findAdvisorState, stockPriceState);
```

Let's see what the state transitions from GetAccountBalance state are:

```
stateMachine.addTransition("accountBalanceIntent",
getAccountBalanceState, getAccountBalanceState,
haveAccTypeShield);
stateMachine.addTransition("accountBalanceIntent",
getAccountBalanceState, getAccTypeState);
stateMachine.addTransition("askForQuoteIntent",
getAccountBalanceState, askforQuoteState);
stateMachine.addTransition("marketTrendIntent",
getAccountBalanceState, marketTrendState);
stateMachine.addTransition("findAdvisorIntent",
getAccountBalanceState, findAdvisorState);
stateMachine.addTransition("stockPriceIntent",
getAccountBalanceState, stockPriceState);
stateMachine.addTransition("weatherIntent",
getAccountBalanceState, getWeatherState);
stateMachine.addTransition("claimStatusIntent",
getAccountBalanceState, getClaimStatusState,
haveClaimIdShield);
stateMachine.addTransition("claimStatusIntent",
getAccountBalanceState, getClaimIdState);
```

Similarly, we can create transitions for other states.

However, note that there is a difference in state transitions defined by GetAccountBalanceState and FindAdvisorState. You can go to GeneralQueryState from FindAdvisorState but not from GetAccountBalanceState. This is where we define which transitions are possible from each state. In the example here, we don't want users to be asking general questions after they ask for account balance details.

A better explanation of this is when the user wants to know his account balance. IRIS will prompt the user on whether he/she wants to know the retirement account balance or annuities account balance and should be expecting a response like one of the following:

- *I am looking for a retirement account balance*

- *Retirement*

- *401k balance*

- *Annuities balance*

- *Want to know 401k account balance*

- *401k*

- *annuities*

Now, in responses to "annuities" or "401k," it is difficult to understand whether the intent is to respond to the question asked or if the user switched the context and is asking something very general that can be searched by IRIS in its knowledge base. For example, a user can also respond with something like

- *What's the weather in Dublin*

- *My claim ID is abc123 can you tell the claim status*

- *Insurance*

- *401k*

- *Retirement funds*

Now, contextually, it is difficult to differentiate between a general query search vs. a response to an account type. Here, we can decide that we will not allow a transition to GeneralQueryState.

Another question is how to understand if it is a general query ask. A general query is never an intent. If no other intent, such as asking for weather details, stock price, market trend, claim status, etc., is applicable and the intent classification engine is not able to classify the user utterance into any of these intent categories with high probability, we by default switch it to a general query.

At this stage, we are done with implementing the IrisConfiguration class, and we have defined intents, matchers, slot, slot types, states, different state transitions, and shields in this class.

Managing State

We now need a helper layer that holds this configuration and seamlessly performs intent matching and then passes this information to trigger state machine actions. This helper layer is StateMachineManager:

```java
public class StateMachineManager {

    /** The intent matcher service for IRIS bot. */
    protected IntentMatcherService intentMatcherService;

    /** The state machine */
    protected StateMachine stateMachine;

    /**
     * Constructs the bot by passing a configuration class that
       sets up the intent matcher service and state machine.
     *
     */
    public StateMachineManager(IrisConfiguration configuration) {
        intentMatcherService = configuration.
        getIntentMatcherService();
        stateMachine = configuration.getStateMachine();
    }

    public String respond(Session session, String utterance)
    throws Exception {
        try {
```

```
        /*
         * Invokes the intentMatcherService.match
           method that returns matched intent.
         * This method sends the user utterance and
           session as an input and obtains matched
           intent from the intent classification
           service.
         */
        MatchedIntent matchedIntent =
        intentMatcherService.match(utterance, session);
        /*
         * This method sends the matched intent as
           an input along with session and gets the
           response back from the state machine.
         */
        String response = stateMachine.
        trigger(matchedIntent, session);
        // The response is returned.
        return response;
    } catch (IllegalStateException e) {
        throw new Exception("Hit illegal state", e);
    }
}
```

At this point, we have IRIS ready for the insurance industry. However, to make it functional, we expose it as a REST service. We need to create a ConversationRequest, a ConversationResponse, a ConversationService, and a ConversationController.

Exposing a REST Service

IRIS is exposed as a REST service and accepts HTTP GET requests. The following is an example of the service running on localhost on port 8080 to accept HTTP GET requests:

```
http://localhost:8080/respond?sender=sender-id&message=user-
message
```

This is a JSON representation:

```
{"message":"response-message-from-service"}
```

ConversationRequest

We create a sample ConversationRequest class in the com.iris.bot. request package that a front-end client can use to send a request to the IRIS bot engine back end via an integration module. In the example client, we integrate with Facebook Messenger. Facebook Messenger provides a sender id, which is a unique user id. This single user identifier helps in creating and maintaining sessions for the user and storing long-term and short-term attributes in memory.

```java
public class ConversationRequest {
    /**
     * Sender id as per Facebook.
     */
    private String sender;

    /**
     * Actual text message.
     */
    private String message;
```

```
/**
 * Timestamp sent by Facebook.
 */
private Long timestamp;

/**
 * Sequence number of the message.
 */
private Long seq;
}
```

ConversationResponse

ConversationResponse is created in the com.iris.bot.response package. The response from the IRIS bot is sent to the integration module through an object of this class.

```
public class ConversationResponse {
    /**
     * Actual reply from the bot
     */
    private String message;
}
```

ConversationService

ConversationService creates a static instance of the state machine manager, which is passed to IrisConfiguration in constructor arguments. It further creates a static instance of the SessionStorage class. These classes are created as static because only one instance of this class should be instantiated. Further, the singleton design pattern could also be used to design these single instances. ConversationService calls a respond method of StateMachineManager and returns a response to the controller.

The controller calls a getResponse method of ConversationService by passing the ConversationRequest.

```
public class ConversationService {

    private static StateMachineManager
    irisStateMachineManager = new StateMachineManager(new
    IrisConfiguration());

    private static SessionStorage userSessionStorage = new
    SessionStorage();

    public ConversationResponse
    getResponse(ConversationRequest req) {
// Default response to be sent if there is a server side exception.
        String response = "Umm...I apologise. Either I am
        not yet trained to answer that or I think I have
        had a lot of Guinness today. "
                    + "I am unable to answer that at the
                    moment. " + "Could you try asking
                    something else Please !";

// If the request message is a salutation like hi or hello,
then instead of passing this information to statemachine
manager, a hard-coded response to salutation can be returned
from the service layer.
        if (req.getMessage().equalsIgnoreCase("hi") || req.
        getMessage().equalsIgnoreCase("hey iris")) {
                response = "Hi There! My name is IRIS (isn't
                it nice! My creators gave me this name). I
                am here to help you answer your queries, get
                you the status of your claims,"
```

```
                          + " tell you about stock prices,
                          find you a financial advisor,
                          inform you about current market
                          trends, help you check your life
                          insurance eligibility "
                          + "or provide you your account
                          balance information.\n"
                          + "Hey, you know what, I can
                          also tell you about current
                          weather in your city. Try asking
                          me out ! ";
        }

// Gets the session object for the sender of the request.
        Session session = userSessionStorage.
        getOrCreateSession(req.getSender());

// Creates a response object.
        ConversationResponse conversationResponse = new
        ConversationResponse();

        try {
// Calls the respond method of state manager by passing session
and message (user utterance).
                response = irisStateMachineManager.
                respond(session, req.getMessage());
// Response is set to the the conversationResponse and returned
to the controller.
                conversationResponse.setMessage(response);
        } catch (Exception e) {
                conversationResponse.setMessage(response);
        }
```

```
        return conversationResponse;
    }
}
```

ConversationController

Finally, there's the controller that exposes `ConversationService` as a REST API by creating an endpoint `/respond`. The implementation is straightforward: the controller receives a GET request, it passes to the service, and the service responds with the response message.

Adding a Service Endpoint

Let's create REST service endpoint using Spring Boot. In Spring's approach to building RESTful web services, HTTP requests are handled by a controller. These components are easily identified by the `@RestController` annotation. The `@RequestMapping` annotation ensures that HTTP requests to `/respond` are mapped to the getKeywordresults() method.

More on how to build a RESTful web service using Java and Spring can be found at `https://spring.io/guides/gs/rest-service/`.

```
@RestController
public class ConversationController {

    @Autowired
    ConversationService conversationService;

    @RequestMapping(value = "/respond", method = RequestMethod.
    GET, produces = MediaType.APPLICATION_JSON_VALUE)
    @ResponseBody
    public ConversationResponse getKeywordresults
    (@ModelAttribute ConversationRequest request) {
```

```
        return conversationService.getResponse(request);
    }
}
```

If we run this on localhost, a sample GET request will be

```
http://localhost:8080/respond?sender=SENDER_ID&message=USER_
MESSAGE&timestamp=TIMESTAMP&seq=MESSAGE_SEQUENCE
```

We create attributes of `ConversationRequest` based on attributes that are sent by Facebook. Hence in the request structure we have `timestamp` and `seq`. However, we do not make use of these two attributes in the demo implementation for intent classification or state transition. Note that these attributes of Messenger webhook events may change with new versions of the Facebook API and can be used in your code depending on your requirements.

Summary

Let's summarize what we discussed in this chapter. We started with the idea of building a basic chatbot framework and why a custom designed chatbot is a needed for the enterprise. Then we discussed the core components of the framework.

First, we discussed intents, utterances, and slots, and defined a custom intent and slot matcher. We also created the `MatchedSlot` and `MatchedIntent` classes.

Then we discussed IRIS memory and how the session can be used to store attributes for the long term and the short term. We discussed the `Session` and `SessionStorage` classes.

We then discussed how a conversation can be modeled as a state machine problem. We discussed the different components of a state machine such as states, transitions, shields, and the `StateMachine` backbone class.

Then we discussed an insurance-focused use case capable of performing certain actions based on different intents and states. We defined various intents, slots, and slot types for the use case. We added these definitions to the configuration class.

We then discussed all the possible states for the use case and explained the execution part of all of these states. As some of the states require a validator before transition, we discussed shields that are required for our example use case. We briefly talked about the general query state and how to leverage an enterprise search in case the utterance is not classified into any of the explicit intents and does not match any document in the knowledge repository.

We then described possible transitions from one state to another depending on the user intent.

We then discussed `StateMachineManager`, which uses the configuration and performs intent matching before triggering state actions.

Lastly, we discussed how to make IRIS functional. We briefly explained how to expose IRIS as a REST service by the creation of service and controller layers.

In the next chapter, we will discuss the other chatbot frameworks available in the marketplace such as RASA, Google Dialogflow, and Microsoft Bot Framework. These frameworks, unlike our build-from-scratch approach, provide many plug-and-play features and make development faster. However, we recommend that you understand the requirements of your enterprise thoroughly before making a choice between the available frameworks.

CHAPTER 7

Introduction to Microsoft Bot, RASA, and Google Dialogflow

In the previous chapter, we discussed how to build an in-house chatbot framework with natural language and conversation capabilities. Building a solution from scratch has advantages that we discussed previously. However, there are use cases and scope where it could be easier, quicker, and cheaper to use readily available online intent classification and conversation management frameworks to build your chatbot client.

In this chapter, we will introduce some of the popular online chatbot frameworks and how to use them.

Microsoft Bot Framework

Microsoft Bot Framework is a comprehensive framework for building enterprise-grade conversational AI experiences. It offers a set of cognitive services through its cloud hosting service called Azure. Azure Cognitive Services enable us to build intelligent enterprise-grade bots.

© Abhishek Singh, Karthik Ramasubramanian, Shrey Shivam 2019
A. Singh et al., *Building an Enterprise Chatbot*,
https://doi.org/10.1007/978-1-4842-5034-1_7

We can use its QnA Maker service to build a quick FAQ bot or use LUIS (Language Understanding Intelligent Service) to build a sophisticated virtual assistant. Figure 7-1 shows the various cognitive services of the bot framework and other components required for an end-to-end conversational AI experience.

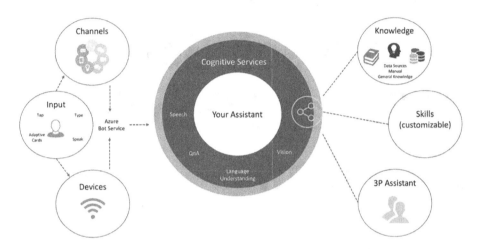

Figure 7-1. *Components of a bot framework*

Introduction to QnA Maker

Microsoft's QnA Maker provides a quick way of building a bot on FAQ URLs, structured documents, manuals, and soon. Its powerful crawler and parser engine extracts all possible questions and answers from the content and makes them available via bot. It is a cognitive service tool that builds and trains a simple QnA bot.

QnA Maker is a free service that responds to a user's utterances or asks in a natural conversational way. It is able to provide helpful answers to semi-structured data with its question/answer mechanism based on its language processing techniques.

A common challenge for most informational bot scenarios is to separate out the content management from the bot design and development, since content owners are usually domain experts who may not be technical. QnA Maker addresses this by enabling a no-code QnA management experience.

QnA Maker allows you to edit, remove, or add QnA pairs with an easy-to-use interface and then publish your knowledge base as an API endpoint for a bot service. It's simple to train the bot using a familiar chat interface, and the active learning feature automatically learns question variations from users over time and adds them to the knowledge base. Use the QnA Maker endpoint to seamlessly integrate with other APIs like a language understanding service and speech APIs to interpret and answer user questions in different ways.

We discussed previously that having a knowledge base is essential to answer users' queries. It is the underlying information repository that provides answers to users' queries. A knowledge base is a logical term for various types of structured, semi-structured, and unstructured information that can be stored in an underlying database with test processing capabilities. QnA Maker exposes its knowledge base/database through API services.

Once we create an account and log in to `www.qnamaker.ai`, it will ask us to create a new knowledge base.

The first step is to create a new Azure QnA service for our knowledge base, which requires creation of the following in Azure:

1. **Creation of a resource group**: Some of the services of QnA Maker are not available in all regions at this point, so we used the (US) west US region for creating and deploying all the components for the demo, as seen in Figure 7-2.

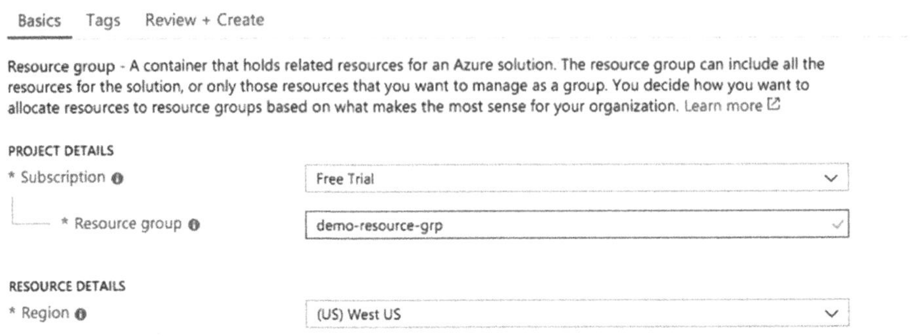

Basics   Tags   Review + Create

Resource group - A container that holds related resources for an Azure solution. The resource group can include all the resources for the solution, or only those resources that you want to manage as a group. You decide how you want to allocate resources to resource groups based on what makes the most sense for your organization. Learn more ☒

PROJECT DETAILS

* Subscription ❶	Free Trial ⌄
⌐— * Resource group ❶	demo-resource-grp ⌄

RESOURCE DETAILS

* Region ❶	(US) West US ⌄

Figure 7-2. *Resource group creation step*

2. **Creation of a QnA Maker resource**: As see in Figure 7-3, we need to provide a name and other details for the resource creation. Please note that we used the Azure Free Tier subscription for the demo. Pricing tiers and subscriptions may vary based on your account.

Figure 7-3. *QnA Maker creation*

Once the QnA Maker resource is created, different resource components will be visible in the Azure All Resources dashboard, as shown in Figure 7-4.

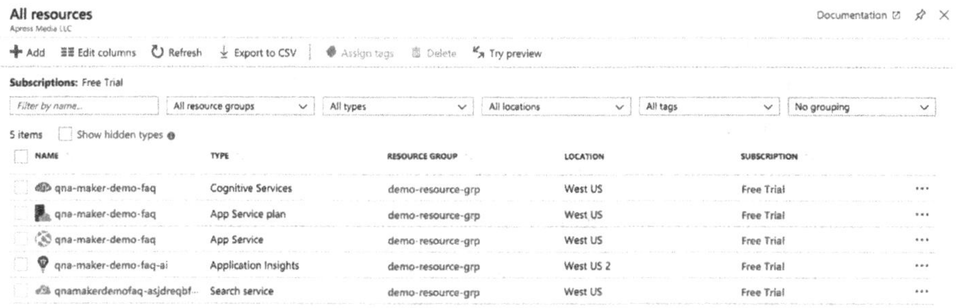

Figure 7-4. *The All Resources dashboard*

Now, we log into `www.qnamaker.ai/Create` and follow the five steps required to create a knowledge base:

1. Create a QnA service in Microsoft Azure (which we just did).

2. Connect the QnA service to the knowledge base.

    As shown in Figure 7-5, you can select the previously created Azure QnA service upon refreshing this page.

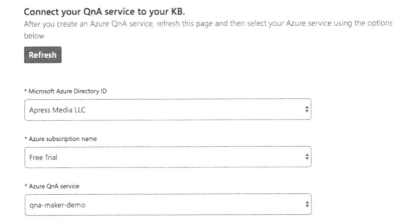

Figure 7-5. *Connecting the QnA service to the knowledge base*

3. Name your knowledge base.

4. Populate the knowledge repository.

    We found a FAQ page on the website of Prudential Financial, a fortune 500 insurance company. The FAQ page had 73 questions and answers that we could use for the demo. This information will be added to the knowledge repository, as shown in Figure 7-6. The reference page URL used for the demo is `www.prudential.com/faq`.

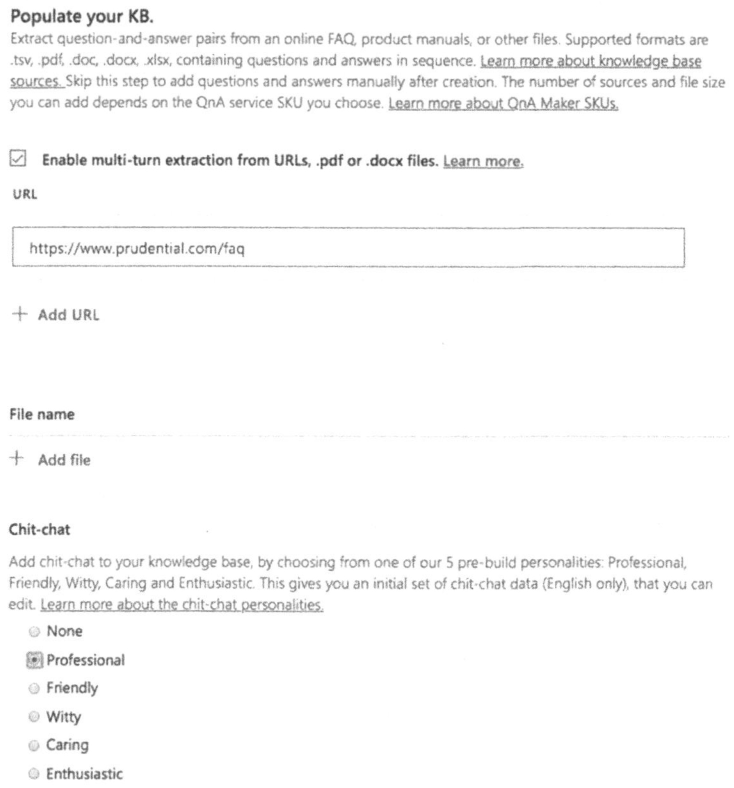

Populate your KB.

Extract question-and-answer pairs from an online FAQ, product manuals, or other files. Supported formats are .tsv, .pdf, .doc, .docx, .xlsx, containing questions and answers in sequence. Learn more about knowledge base sources. Skip this step to add questions and answers manually after creation. The number of sources and file size you can add depends on the QnA service SKU you choose. Learn more about QnA Maker SKUs.

☑   **Enable multi-turn extraction from URLs, .pdf or .docx files.** Learn more.

URL

https://www.prudential.com/faq

\+   Add URL

File name

\+   Add file

Chit-chat

Add chit-chat to your knowledge base, by choosing from one of our 5 pre-build personalities: Professional, Friendly, Witty, Caring and Enthusiastic. This gives you an initial set of chit-chat data (English only), that you can edit. Learn more about the chit-chat personalities.

○ None
◉ Professional
○ Friendly
○ Witty
○ Caring
○ Enthusiastic

Figure 7-6.   *Populating the knowledge base with data from a URL source*

5.   Create the knowledge base.

The QnA maker service crawls the provided URL and extracts information in the form of questions and answers for us to review, add more data, and save and train, as shown in Figure 7-7. There are more options, such as testing the service before publishing.

Figure 7-7.   Extraction of question/answer pairs by the QnA service

The next step is to publish this knowledge base for use by the bot by clicking the Publish button, as shown in Figure 7-8.

Figure 7-8.   Publishing the knowledge repository

Upon publishing the knowledge base, the previously created qna-maker-demo QnA service is ready to be used. We can then make a POST request to the service to query the knowledge base with a question and it will respond with an answer. Figures 7-9 and 7-10 show an example request via POSTMAN for the question asked ("change beneficiary"), which the QnA service matched with a similar question in the knowledge base ("How do I change my beneficiary?").

Figure 7-9. *HTTP POST request to demo-faq-kb knowledge base with authorization*

Figure 7-10. *The question asked for the response shown in Figure 7-9*

You can see how to use Microsoft QnA Maker using Azure to quickly build a fully functional Q&A service. This service can be integrated with a WebApp bot client or Facebook Messenger or any other third-party messaging service.

The functionality of the QnA Maker service is essentially a relevant search engine in the back end that indexes content data and uses a text-based scoring logic to rank best matching content. However, this service is not conversational and does not perform language understanding to identify user intents before taking an action. Microsoft provides another cognitive service called LUIS that is created for this purpose.

Introduction to LUIS

Language Understanding Intelligent Service (LUIS) is a cloud-based API service and a component of cognitive services, as shown in Figure 7-1. It takes user input and applies machine learning techniques to predict user intent. It also uses its Named Entity Recognition (NER) matching techniques to identify named entities in the user utterance to provide a meaningful response. This is illustrated in Figure 7-11. In a nutshell, LUIS is an intent classification engine that provides the best matching (top scoring) intent based on the user utterance.

The first step in the process is to create a LUIS app consisting of a domain-specific language model. At the time of creation of the app, we can select a prebuilt domain model, build our own natural language model, or leverage the prebuilt domain with our own customization.

- **Prebuilt model**: LUIS provides many prebuilt domain models including predefined intents of common usage with various utterances and prebuilt entities of various domains. These prebuilt entities can also be leveraged without having to use intents and utterances of the prebuilt model. Depending on whether the prebuilt model is suitable for your use case, prebuilt models can be a good, quick starter.

- **Custom entities**: LUIS provides several ways in which we can define our custom intents along with sample utterances. We can also define our domain-centric custom entities leveraging NER matching capabilities.

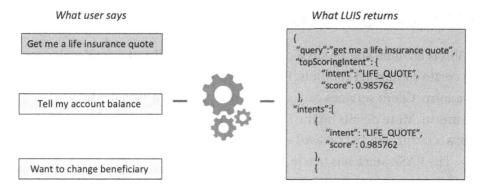

Figure 7-11. *LUIS intent classification illustration*

The creation of a LUIS app is simple and similar to what we will discuss when we create an Alexa app (for integration with IRIS) in Chapter 9, so we will skip the explanation of the same in this chapter.

Once the intents, utterances, and domain-related entities are defined using the custom or prebuilt model, the LUIS app is then published. Publishing an app means making it available for use. Once published, a client application can send utterances to a LUIS service endpoint, as illustrated in Figure 7-12. LUIS responds with a JSON response.

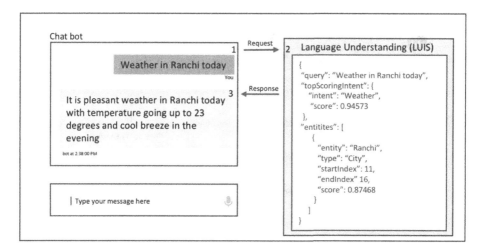

Figure 7-12. *Example LUIS bot flow*

Introduction to RASA

RASA is an open source stack of machine learning libraries for developers to create contextual chatbots. RASA also provides a paid enterprise-grade platform. Client services can connect with various APIs in the RASA platform. More details on the RASA platform can be found at `https://rasa.com/products/rasa-platform/`.

The RASA stack has two main components:

- **Core**: A chatbot framework similar to Mutters, with machine learning-based dialog management

- **NLU**: An intent classification module providing intent classification and entity extraction.

As shown in Figure 7-13, both Core and NLU modules provide various functionalities that can be customized, trained, and integrated on a custom dataset for contextual AI-based virtual assistants.

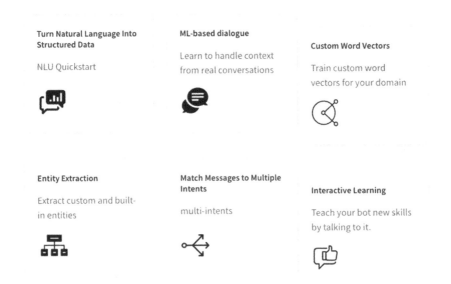

Figure 7-13. *Functionalities of Core and NLU modules*

- **Turn natural language into structured data**: This functionality is similar to what we discussed in the QnA Maker example, in which structured, semi-structured, and unstructured data was parsed in some meaningful form in the knowledge base for querying.

- **ML-based dialogue:** RASA provides its own dialogue management module that can be customized and modelled via a series of simple steps. We discussed an implementation of conversation management in IRIS in Chapter 6 with finite state machines.

- **Custom word vectors**: RASA provides customization to train word vectors for our domain. Several word embeddings are generated in the process, which helps the engine in better intent classification. Word2vec is a popular neural network model that generates word embeddings from user text.

- **Entity extraction**: By using the built-in or custom-built entity recognition models, RASA can identify entities in user utterances.

- **Match messages to multiple intents**: A user utterance can be matched to multiple intents, and different match scores along with the best matching intent are returned in response.

- **Interactive learning**: We discussed in the last chapter how to enhance IRIS by extending the design to support a continuous improvement framework through interactive learning. RASA also provides simple ways to enable interactive learning that teaches the bot new skills.

The RASA framework for chatbots makes use of the two models to effectively flow the conversation, as shown in Figure 7-14.

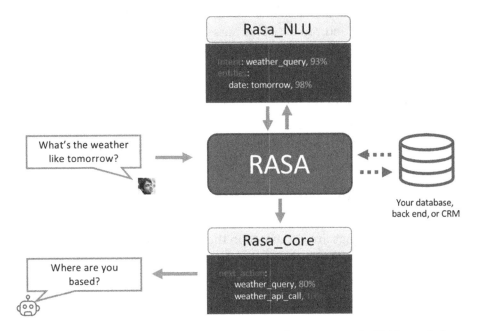

Figure 7-14. *Flow of conversation using the Core and NLU modules of RASA*

The message is basically handled at two stages: one at the Core module, which manages the conversation, and one at the NLU module, which provides the essential language services. The external API calls or CRM connections are dealt with by the RASA platform with the help of RASA core trackers. The next section discusses some more details on how the two modules work.

RASA Core

RASA Core refers to the main component, which receives and responds to the requests. The module is robust and has a flow-based approach to handle all requests, as shown in Figure 7-15.

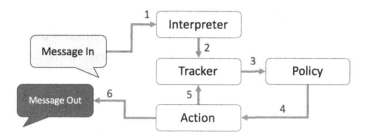

Figure 7-15. *Working on the RASA framework*

The RASA framework has a six-step flow to handle all incoming messaging requests and respond to them. The steps are described below; however, more details can be found at `http://rasa.com/docs/#rasa_core.agent.Agent`.

The interpreter receives the message and converts it into a dictionary including the original text, the intent, and any entities that were found. The tracker keeps track of the conversations and passes on the state with new message to policy. The policy module prepares the response for the request. Once the response is ready in the policy module, it is passed to the tracker and action. The tracker updates the state of conversation and the action sends the response back to the user.

The above implementation helps all the conversations to flow through RASA and at the same time keeps track of conversations to maintain states of conversations.

RASA NLU

RASA Natural Language Understanding (NLU) module is a tool that does the intent classification and entity extraction for all the messages incoming from the user. It is a fully open source tool developed for the core purpose of being used in chatbot development.

The RASA NLU can be hosted as a service for your chatbot to use. The RASA NLU also allows you to train different types of models using your own data. From the perspective of chatbot developers, the RASA NLU module has two parts:

- **Training**: This module allows you to train models on your own data. Having your own data to train allows you to develop a NLU that is business specific.

- **Server**: This module serves as the training model to the chatbot. It can be hosted as an API service and runs at the back end.

The key benefit of using the open source NLU module is that you don't need to send your data outside to Google or Amazon to train on intents and entities. And, being open source, you can tweak the models or develop your own models as per your needs. The architecture for the RASA NLU also allows it to run from everywhere as a service. You may not need to make network calls if your architecture demands that.

Refer to the extensive documentation of RASA at `https://rasa.com/docs/`.

Introduction to Dialogflow

Dialogflow is an offering of Google for developing human-computer interactions. It was formerly known as API.ai as a platform for managing APIs for chatbots. Google bought the company in September 2016 and renamed it Dialogflow in 2017. It is now a core offering on Google for voice-based application development for all of its platforms including Google Home, Google Assistant, and other voice-enabled services. The capability of Dialogflow to integrate with virtually all platforms including wearables, home applications, speakers, smart devices, etc. makes it one of the most popular platforms for developing chatbots.

Google Dialogflow functioning is similar to the previous discussed chatbots, with some variations in naming technology and technical architecture on how it implements the flow. A typical conversation with a chatbot in Dialogflow looks like Figure 7-16.

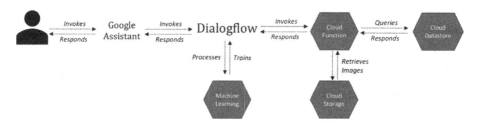

Figure 7-16. *Dialogflow-based chatbot*

The user responses are allowed in both written format and in voice/ speech. The responses are captured by the Google Assistance module and then converted into text to allow the chatbot to be voice enabled. Google Assistance can be replaced by other integration options as well.

The key steps to building a chatbot with Dialogflow are discussed below. You are encouraged to go through the official documentation for Dialogflow at `https://dialogflow.com/`.

- The first step is to create a Dialogflow account and then create a Dialogflow agent, which lets you define a natural language understanding model.

- The second step is to define how information is extracted from user utterances. This is done by defining extract parameters for entities. This is helpful to understand important attributes in user utterances, providing better matches based on extracted entities.

- In order for a chatbot to be conversational, we need to define and manage states with contexts in the next step.

- Finally, the Dialogflow agent can be integrated with Google Assistant, which lets us deploy the agent. This allows users to invoke actions through the assistant to interact.

The above tutorial details can be found at `https://dialogflow.com/docs`. Below we present selected steps in the process to build a chatbot using Dialogflow.

Creating an agent means creating an NLU model that will understand the chats happening with that agent. So, if you want to create a chatbot for two different domains, you may need to create two different agents and train them accordingly. See Figure 7-17.

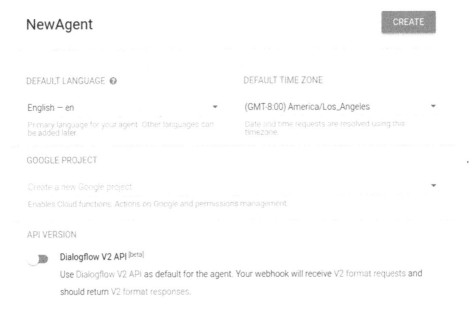

Figure 7-17. Creation of an agent in Dialogflow

The next step is to create intents using the Dialogflow console. The intent can be defined along with the utterances using the rich UI of Dialogflow. The console window looks similar to Figure 7-18.

Figure 7-18. *Intent creation in Dialogflow*

The Dialogflow console also provides a simulator at the right side of the window to test the trained intents in a simulated environment. This helps you to test your model before deploying it to production. The simulation window looks similar to Figure 7-19.

What's your name?

See how it works in Google Assistant.

Agent   Domains

USER SAYS   COPY CURL
What's your name?

DEFAULT RESPONSE   ▼   PLAY
My name is Dialogflow!

INTENT
Name

ACTION
Not available

SHOW JSON

Figure 7-19. *Dialogflow simulation window*

Actions and parameters are extracted from the entity extraction module of Dialogflow. The parameters extracted then can be used for fulfilment and generating the responses. The Entity extraction mechanism is driven by powerful Google ML models. A basic training phase and entity extraction in Dialogflow is shown in Figure 7-20.

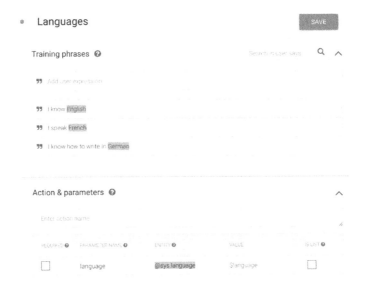

Figure 7-20. *Entity extraction in Dialogflow*

The contexts can be managed by the console as well, and your bot can be trained on missed intents that the chatbot was not able to understand. Every time you simulate the model, it gets trained on the latest data provide to its intent engine.

Dialogflow connects with outside services and CRMs using its fulfilment service. The fulfillment service allows you to create custom API endpoints and expose them to your chatbot. The fulfillment service in Dialogflow is called Web Hooks.

One of the great features is the capability to integrate the chatbot with multiple platforms with just few clicks. Figure 7-21 shows the integration options available with Dialogflow. The ecosystem of Dialogflow is very strong, and it offers good integration models in Nodejs for developers to build chatbots and features as per their requirements.

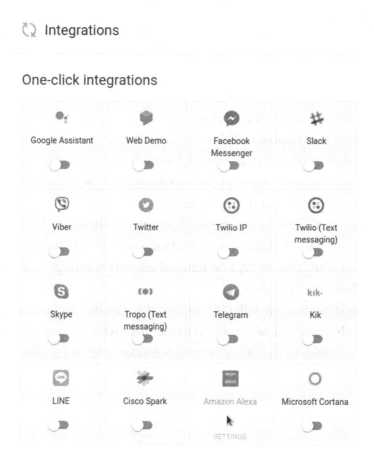

Figure 7-21.   *Integration options in Dialogflow*

Summary

We discussed different chatbot frameworks and how they are useful to build a simple chatbot client. However, these frameworks have advantages and disadvantages.

Microsoft Bot Framework is open source and available on GitHub but some of its services such as Lex or Azure come at an additional cost. It's designed to be easy to use as a set of services.

RASA is also open source and provides enterprise platform support. It requires development and programming knowledge since it is designed for developers. It does not provide out-of-the-box integration to other messaging clients. It can be hosted on-premise, but it does not provide direct user info management support.

Dialogflow provides out-of-the-box integration with some popular messaging platforms such as Facebook Messenger, Skype, and Google Assistant and provides the complete toolkit required for building bots. However, it cannot be operated on-premise.

The chapter also discussed the key features of various platforms and presented essential information required for developers to start working on these platforms. The chapter concluded with an introduction to an enterprise-level paid platform.

In the next chapter, we will discuss how the custom chatbot designed in previous chapters can be extended to integrate with various third-party services and enterprise databases to provide responses to the client.

CHAPTER 8

Chatbot Integration Mechanism

In Chapter 6, we designed a simple chatbot framework in Java which we called the IRIS (Intent Recognition and Information Service) framework. We discussed the core components of IRIS, such as how to define intents and how state machine can be implemented for defining state and transitions for building a conversational chatbot. An example use case focused on the insurance domain. In the example, we outlined specific capabilities that IRIS is supposed to perform such as providing market trend details, stock price information, weather details, and claim status.

In this chapter, we will focus on the integration modules of IRIS, which shows how we can connect with external data sources and third-party APIs for information retrieval.

Integration with Third-Party APIs

In our example, the three functionalities of IRIS require integration with third-party APIs (Figure 8-1):

- Market trends

- Stock prices

- Weather information

© Abhishek Singh, Karthik Ramasubramanian, Shrey Shivam 2019
A. Singh et al., *Building an Enterprise Chatbot*,
https://doi.org/10.1007/978-1-4842-5034-1_8

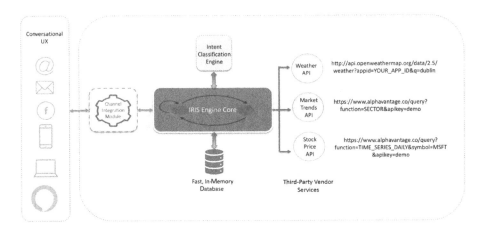

Figure 8-1. *IRIS integration with third-party APIs*

Market Trends

There are plenty of free and paid APIs available online that can provide these details. We explored www.alphavantage.co, which provides free APIs to get real-time and historical stock market data. Alpha Vantage APIs are grouped into four categories:

- Stock time series data

- Physical and digital/crypto currencies (e.g., Bitcoin)

- Technical indicators

- Sector performances

All APIs are in real time: the latest data points are derived from the current trading day.

Upon providing just three necessary details, which are the type of user, institution/organization name, and an email address, we get an API key, and it is free for a lifetime as per the website. The free API key can be obtained by providing details at www.alphavantage.co/support/#api-key.

Once we have the API key, there are various APIs of Alpha Vantage grouped under multiple API suites, as shown in Figure 8-2.

- Stock time series

- Forex

- Technical indicators

- Cryptocurrencies

- Sector performances

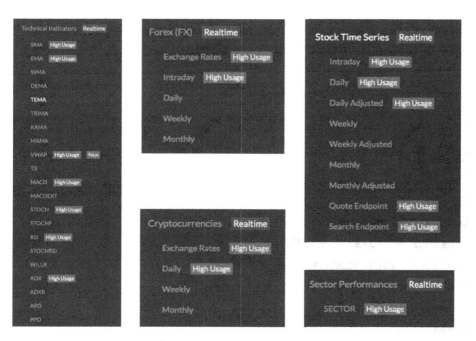

Figure 8-2. *Alpha Vantage's multiple API suites*

More details on each of these APIs can be found at `www.alphavantage.co/documentation/`.

For our example use case, we want to know how to get the current market trend and stock price of a particular stock. For the current market trend, we leverage the Sector Performances API, details of which are available at `www.alphavantage.co/documentation/#sector-information`.

A sample HTTP GET request to obtain real-time, sector-wise performance details is at `www.alphavantage.co/query?function=SECTOR&apikey=demo`.

The JSON response received from the API is

```
{
Meta Data:
{
Information: "US Sector Performance (realtime & historical)",
Last Refreshed: "03:44 PM ET 03/04/2019"
},
Rank A: Real-Time Performance:
{
Real Estate: "0.47%",
Materials: "0.39%",
Utilities: "0.11%",
Energy: "-0.10%",
Communication Services: "-0.10%",
Consumer Staples: "-0.22%",
Consumer Discretionary: "-0.23%",
Industrials: "-0.33%",
Information Technology: "-0.50%",
Financials: "-0.65%",
Health Care: "-1.49%"
},
Rank B: 1 Day Performance:
{
Energy: "1.81%",
Health Care: "1.41%",
Consumer Discretionary: "0.92%",
Communication Services: "0.78%",
Information Technology: "0.71%",
```

```
Financials: "0.54%",
Utilities: "0.19%",
Industrials: "0.09%",
Real Estate: "-0.13%",
Materials: "-0.16%",
Consumer Staples: "-0.17%"
},
Rank C: 5 Day Performance:
{},
Rank D: 1 Month Performance:
{},
Rank E: 3 Month Performance:
{},
Rank F: Year-to-Date (YTD) Performance:
{},
Rank G: 1 Year Performance:
{},
Rank H: 3 Year Performance:
{},
Rank I: 5 Year Performance:
{},
Rank J: 10 Year Performance:
{}
}
```

In this response, some data points are available such as real-time performance, one day, five days, and one-month performance. The **Real-Time Performance** key in the JSON response describes the different sectors of the market along with their percentage change in real time. We use this information to provide current market trend information in the MarketTrendState:

```java
public String execute(MatchedIntent matchedIntent, Session
session) {
```

```java
// Third-party API that provides us the current market trend
String uri = "https://www.alphavantage.co/query?function=SECTOR
&apikey=YOUR_KEY";
// Java client that performs HTTP request and gets a response
by performing a GET call to the URL
            RestTemplate restTemplate = new RestTemplate();
/*
 * Response is mapped to a string object below. However, in
   actual development, we should create a Java Bean (POJO) that
   will be used to map the response into a Java response object
   by using the getForObject method.
 */
            String result = restTemplate.getForObject(uri,
            String.class);
            String answer = "";

/*
 * ObjectMapper provides functionality for reading and writing
   JSON, either to and from basic POJO. It is part of Jackson,
   a standard Java library for parsing JSON.
 */
            ObjectMapper mapper = new ObjectMapper();
            try {

/*
 * JsonNode is used to parse response in a JSON tree model
   representation by Jackson. JsonNode is a base class for all
   JSON nodes, which form the basis of JSON Tree Model that
   Jackson implements. One way to think of these nodes is to
   consider them similar to DOM nodes in XML DOM trees
```

```
*/

        JsonNode actualObj = mapper.
        readTree(result);
        JsonNode jsonNode1 = actualObj.get("Rank A:
        Real-Time Performance");

        answer = "Energy Sector is " + jsonNode1.
        get("Energy").textValue() + ". Utilities at "
                + jsonNode1.get("Utilities").
                textValue() + ". Real Estate at "
                + jsonNode1.get("Real Estate").
                textValue() + ". Consumer
                Staples at "
                + jsonNode1.get("Consumer
                Staples").textValue() + ".
                Health Care at "
                + jsonNode1.get("Health Care").
                textValue() + ". Materials at "
                + jsonNode1.get("Materials").
                textValue() + ".
                Telecommunication Services at "
                + jsonNode1.
                get("Telecommunication
                Services").textValue() + ".
                Industrials at "
                + jsonNode1.get("Industrials").
                textValue() + ". Information
                Technology at "
                + jsonNode1.get("Information
                Technology").textValue() + ".
                Consumer Discretionary at "
```

```
                              + jsonNode1.get("Consumer
                              Discretionary").textValue() +
                              ". Financials at "
                              + jsonNode1.get("Financials").
                              textValue() + "\nWhat else do
                              you want to know?";
        } catch (Exception e) {
                e.printStackTrace();
                Result = "I am unable to retrieve this
                information right now. There is some problem
                at my end.\nTry asking something else!";
        }
        return answer;
    }
```

Stock Prices

A user may interact with IRIS and may ask for stock price information. The utterance(s) could be

- What is the current stock price of microsoft
- Pru stock price
- Infy stock today
- Share price of hdfc

When the utterance is received by IRIS Core, it passes the utterance to the intent classification engine, which knows that the user is looking for 'STOCK_PRICE'. Based on this intent and current state, a transition to StockPriceState happens. The execute method of this state then makes a call to the third-party API.

To retrieve the stock price details, we use a TIME_SERIES_DAILY API from the Stock time series API suite of Alpha Vantage.

A sample HTTP GET request:

```
www.alphavantage.co/query?function=TIME_SERIES_DAILY&symbol=MSF
T&apikey=demo
```

A sample API response:

```
{
Meta Data:
{
1. Information: "Daily Prices (open, high, low, close) and
   Volumes",
2. Symbol: "MSFT",
3. Last Refreshed: "2019-03-04 16:00:01",
4. Output Size: "Compact",
5. Time Zone: "US/Eastern"
},
Time Series (Daily):
{
2019-03-04:
{
1. open: "113.0200",
2. high: "113.2000",
3. low: "110.8000",
4. close: "112.2600",
5. volume: "25684300"
},
2019-03-01:
{
1. open: "112.8900",
2. high: "113.0200",
3. low: "111.6650",
```

4. close: "112.5300",
5. volume: "23501169"
},
2019-02-28:
{
1. open: "112.0400",
2. high: "112.8800",
3. low: "111.7300",
4. close: "112.0300",
5. volume: "29083934"
},
2019-02-27:
{},
2019-02-26:
{}
}
}

A sample execute method of StockPriceState could look like:

```java
public String execute(MatchedIntent matchedIntent,
Session session) {

/*
* In the URL below, we have hard-coded symbol=MSFT. MSFT is the
  symbol for Microsoft. In an actual implementation, we should
  retrieve the company name from the user utterance and find
  its symbol and then pass it in the GET request below. There
  are many ways to convert from company name to symbol such as
  by calling publicly available services or by maintaining a
  mapping.
*/
```

```
String uri = "https://www.alphavantage.co/
query?apikey=YOUR_KEY&function=TIME_SERIES_DAILY&o
utputsize=compact&symbol=MSFT";

RestTemplate restTemplate = new RestTemplate();
String result = restTemplate.getForObject(uri,
String.class);
```

// Default answer in case the third-party API does not respond
or if there is any network related issue

```
String answer = "I am somehow unable to retrieve
stock price details right now. But I will be able
to help you with your other queries.";

ObjectMapper mapper = new ObjectMapper();
try {
```

/*
 * As we know, the stock market does not run on the weekends
   and certain holidays. IRIS is expected to provide real-time
   stock performance data. In a normal working day, we parse
   out performance detail of current day but for a holiday or
   if the stock market was closed or if the stock did not trade
   that day, we get the performance detail of the previous day.
 */

```
Date date = new Date();
String today = new SimpleDateFormat("yyyy-
MM-dd").format(date);
String yday = new SimpleDateFormat("yyyy-MM-
dd").format(yesterday(1));
String dayBeforeYday = new
SimpleDateFormat("yyyy-MM-dd").
format(yesterday(2));
```

```java
JsonNode actualObj = mapper.readTree(result);
JsonNode jsonNode1 = actualObj.get("Time
                        Series (Daily)");
JsonNode jsonNode2 = jsonNode1.get(today);
JsonNode jsonNode3 = jsonNode1.get(yday);
JsonNode jsonNode4 = jsonNode1.
                        get(dayBeforeYday);
if (jsonNode2 != null) {
    answer = "Today Microsoft stock opened
    at " + jsonNode2.get("1. open").
    textValue() + " and closed at "
                + jsonNode2.get("4.
                close").textValue();
    answer = answer + " It saw an
    intraday high of " + jsonNode2.
    get("2. high").textValue()
                + " and an intraday low
                of " + jsonNode2.get("3.
                low").textValue();
    answer = answer + ". Total volume
    traded is " + jsonNode2.get("5.
    volume").textValue() + "\n"
                + "How else can I help
                you?";
} else if (jsonNode3 != null) {
    answer = "I don't have the trading
    info for today as of now, but
    Yesterday PRU stock opened at "
                + jsonNode3.get("1.
                open").textValue() + "
                and closed at "
```

```
                    + jsonNode3.get("4.
                    close").textValue();
            answer = answer + " It saw an
            intraday high of " + jsonNode3.
            get("2. high").textValue()
                    + " and an intraday low
                    of " + jsonNode3.get("3.
                    low").textValue();
            answer = answer + ". Total volume
            traded is " + jsonNode3.get("5.
            volume").textValue() + "\n"
                    + "How else can I help
                    you?";
} else if (jsonNode4 != null) {
        answer = "On friday, before weekend,
        PRU stock opened at " + jsonNode4.
        get("1. open").textValue()
                    + " and closed at "
                    + jsonNode4.get("4.
                    close").textValue();
        answer = answer + " It saw an
        intraday high of " + jsonNode4.
        get("2. high").textValue()
                    + " and an intraday low
                    of " + jsonNode4.get("3.
                    low").textValue();
        answer = answer + ". Total volume
        traded is " + jsonNode4.get("5.
        volume").textValue() + "\n"
                    + "How else can I help
                    you?";
```

```
                }
        } catch (Exception e) {
                e.printStackTrace();
        }
        return answer;
    }

    /*
     * A method to return 'days' before the current day. If
       the value of 'days' is 1, yesterday's date is returned.
     * If the value is 2, the day before yesterday is
       returned and so on.
     */
    private Date yesterday(int days) {
        final Calendar cal = Calendar.getInstance();
        cal.add(Calendar.DATE, -days);
        return cal.getTime();
    }
```

Weather Information

There are plenty of digital bots available that provide weather details.
People often ask Siri, Google voice assistant, and Alexa to give details on
the weather. Let's see how we can use a third-party API to integrate with
IRIS for weather information.

To get the weather report, we leverage http://openweathermap.org,
which provides the API to get weather details of the requested city. It offers
multiple data points such as current weather data, 5-day forecast, 16-day
forecast, and other historical information about the city. It currently
includes weather details for over 200,000 cities around the world. The
current weather is frequently updated based on global models and data
from more than 40,000 weather stations. OpenWeather also provides APIs

for relief maps, managing personal weather stations, bulk downloading, weather alerting, UV index, and air pollution.

For our example, we need the current weather in a given city. We need to obtain an API key. OpenWeather provides multiple API plans, the details of which can be found at `https://openweathermap.org/price`.

There is a free plan that allows a maximum of 60 calls per minute, which is more than enough for the demo. We use the Current Weather data API that can be called in multiple ways to get weather details such as the following:

- Call the current weather data for one location:

  - By city name

  - By city ID

  - By geographic coordinates

  - By ZIP code

- Call the current weather data for several cities:

  - Cities within a rectangle zone

  - Cities in cycle

  - Call for several city IDs

A sample HTTP GET request when querying by city name (passed in q):
`http://api.openweathermap.org/data/2.5/weather?appid=YOUR_APP_ID&q=dublin`

The JSON response:

```
{
coord:
{},
weather:
[
```

```
{
id: 501,
main: "Rain",
description: "moderate rain",
icon: "10n"
}
],
base: "stations",
main:
{
temp: 277.07,
pressure: 993,
humidity: 100,
temp_min: 275.93,
temp_max: 278.15
},
visibility: 10000,
wind:
{
speed: 6.2,
deg: 230
},
rain:
{
1h: 1.14
},
clouds:
{
all: 75
},
dt: 1551735821,
```

```
sys:
{
type: 1,
id: 1565,
message: 0.0045,
country: "IE",
sunrise: 1551683064,
sunset: 1551722984
},
id: 2964574,
name: "Dublin",
cod: 200
}
```

A sample execute method of GetWeatherState may contain the following code snippet:

```
public String execute(MatchedIntent matchedIntent, Session
session) {
/*
* Default response in case there is a network issue or if the
  third-party API takes a lot of time or if there is some other
  exception
*/
                String answer = "I am unable to get the weather
                report right now. But I hope it be a nice and
                charming day for you :) ";
                /*
                 * GET API that provides weather details
                 */
                String uri = "http://api.openweathermap.org/
                data/2.5/weather?appid=YOUR_API_KEY&q=";
```

```java
String cityName = "dublin";
try {
     RestTemplate restTemplate = new RestTemplate();
     String result = restTemplate.
     getForObject(uri, String.class);

     ObjectMapper mapper = new ObjectMapper();
     JsonNode actualObj = mapper.readTree(result);
     ArrayNode jsonNode1 = (ArrayNode) actualObj.
     get("weather");
     JsonNode jsonNode2 = actualObj.get("main");

     String description = jsonNode1.get(0).
     get("description").textValue();

     String temperature = jsonNode2.get("temp").
     toString();
     Double tempInCelsius = Double.
     parseDouble(temperature) - 273.15;
     double roundOff = Math.round(tempInCelsius *
     100.0) / 100.0;
     String humidity = jsonNode2.get("humidity").
     toString();

     answer = "It seems to be " + description +
     " at the moment in " + cityName + ". The
     temperature is "
                    + roundOff + " degrees.
                    Humidity" + " is close to " +
                    humidity
                    + ".\n I wish I were human to
                    feel it. Anyways, what else do
                    you want to know from me? ";
```

```
    } catch (Exception e) {
    }
    return answer;
}
```

Connecting to an Enterprise Data Store

In our example, we use GetClaimStatus to demonstrate how to connect to a database and query claims information.

Note that although we are demonstrating this capability of querying a database directly, the modern design approach across the industry does not recommend it. Any database should be queried only through a service created on top of the database. There are multiple reasons for this, such as security and access control, database load and connection handling, encapsulation, and portability.

```
public class GetClaimStatus extends State {
    /*
     * Java Database Connectivity (JDBC) is an application
     *   programming interface (API) for the programming
     *   language Java, which defines how a client may
     *   access a database. It is a Java-based data access
     *   technology used for Java database connectivity. It
     *   is part of the Java Standard Edition platform, from
     *   Oracle Corporation
     * DB_URL is the database connection URL.
     * The URL used is dependent upon the particular
     *   database and JDBC driver. It will always begin with
     *   the "JDBC:" protocol, but the rest is up to the
     *   specific vendor. In our example, we use a MySQL
     *   database.
     */
```

```
static final String JDBC_DRIVER = "com.mysql.jdbc.Driver";
// database name is a test
static final String DB_URL = "jdbc:mysql://localhost/test";

/*
 * Database access credentials
 */
static final String USERNAME = "ClaimsReadOnlyUser";
static final String PASSWORD = "**********";

public GetClaimStatus() {
        super("getClaimStatus");
}

@Override
public String execute(MatchedIntent matchedIntent,
Session session) {
        Connection conn = null;
        Statement stmt = null;
        String status = null;
/*
* Retrieve claim Id from a session or slot of the matched
  intent. If this state is executed, it is supposed to mean
  that we have the claim Id; otherwise Shield would not have
  validated transition to this state.
 */
        String claimId = SessionStorage.getStringFromSlotOr
        Session(matchedIntent, session, "claimId", null);

// Default answer
        String answer = "We dont have any info related to
        " + claimId + " in our system.\n"
```

```
                + "Please contact our call service
                representative for further inquiry
                on the number 1800 333 0333 between
                Mondays to Fridays, 8:30 am to
                5:30 pm.\n"
                + "If you're dialling from overseas
                or via a payphone, please call
                +65 633 30333.\nIs there anything
                else I can help you with?";
        try {
                //Register JDBC driver
                Class.forName("com.mysql.jdbc.Driver");

                // Open a connection (Connecting to
                database...)
                conn = DriverManager.getConnection(DB_URL,
                USERNAME, PASSWORD);

                // Execute a query
                stmt = conn.createStatement();
/*
 * SQL query to query row from test database and claims table.
   This query means - return status of the row from claims table
   where claim id is given claim Id.
 */
                String sql = "SELECT status FROM claims
                where claimId='" + claimId + "'";

                //executing SQL
                ResultSet rs = stmt.executeQuery(sql);

                //Extract data from the result set
                while (rs.next()) {
```

```
                    //record fetched
                    status = rs.getString("status");
            }

            //Clean up environment and close active
            connections.
            rs.close();
            stmt.close();
            conn.close();
        } catch (Exception e) {
            e.printStackTrace();
        } finally {
```
```
/*
 * In a try-catch, a finally block is always executed even if an
   exception happens. In case of exceptions in the above code for
   any reason, the statements and connections will not get closed.
   Hence we apply an extra check to close it in finally block.
 */
```
```
            try {
                if (stmt != null)
                    stmt.close();
            } catch (SQLException se2) {
            }
            try {
                if (conn != null)
                    conn.close();
            } catch (SQLException se) {
                se.printStackTrace();
            }
        }
```

```
// If we received status from the database for that claims, we
override the default answer with actual status details.
            if (status != null) {
                answer = "The status of your claim for
                claimId " + claimId + " is  - " + status
                                + ".\nContact our
                                representatives at HELPLINE-
                                NUMBER "
                                + "between Mondays to Fridays,
                                8:30am to 5:30pm if you want to
                                inquire more. Anything else that
                                you want to know as of now?";
            }
            // Remove claim Id related attributes from the session
            session.removeAttribute("getclaimidprompt");
            session.removeAttribute("claimid");
            return answer;

        }
}
```

Integration Module

The integration module is the piece that connects the core of IRIS with different messaging platforms such as Facebook Messenger, Twitter, the Web, mobile apps, Alexa, etc. Each of these platforms has their own integration mechanism, and building a customized integration layer for each of the channels is very difficult and not the core objective in this book. The integration module is a middle-layer service exposed to the outside world, acting as a gateway for IRIS, as shown in Figure 8-3.

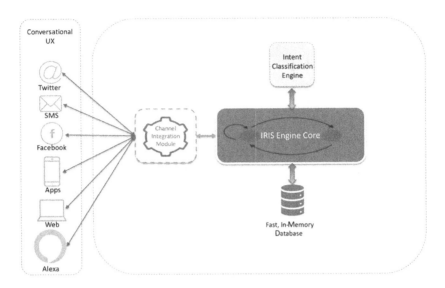

Figure 8-3. *IRIS channel integration module*

Many open source tools provide simple customized integrations with various channels without us having to write a lot of code, and they take care of a lot of complexity required for the integrations. In this illustration, we focus on bootkit, a leading development tool providing customized combinations with multiple messaging platforms:

`https://github.com/howdyai/botkit.`

In this chapter, we will discuss integration of IRIS with Facebook Messenger. Before we get into the integration code, we need to ensure we have the following in order to be able to put IRIS on a Facebook Messenger page:

- **A Facebook page**: A Facebook page is used as the identity of your bot. When people chat with your app, they see the page name and profile picture.

- **A Facebook developer account**: Your developer account is required to create new apps, which are the core of any Facebook integration. You can create a new developer account by going to Facebook for Developers and clicking the Get Started button.

- **Facebook App for Web**: The Facebook app contains the settings for your Messenger bot, including access tokens.

- **A webhook URL**: Actions that take place in conversations with your bot, such as new messages, are sent as events to your webhook. This is the URL of our integration module, which we cover next.

The setup process requires adding the Messenger platform to your Facebook app, configuring the webhook of the app, and subscribing your app to the Facebook page. The details on setting up the Facebook app can be found at `https://developers.facebook.com/docs/messenger-platform/getting-started/app-setup/`.

A step-by-step guide to configuring the botkit and Facebook Messenger is available at `www.botkit.ai/docs/provisioning/facebook_messenger.html`.

1. Create the app and select the Messenger platform. See Figure 8-4.

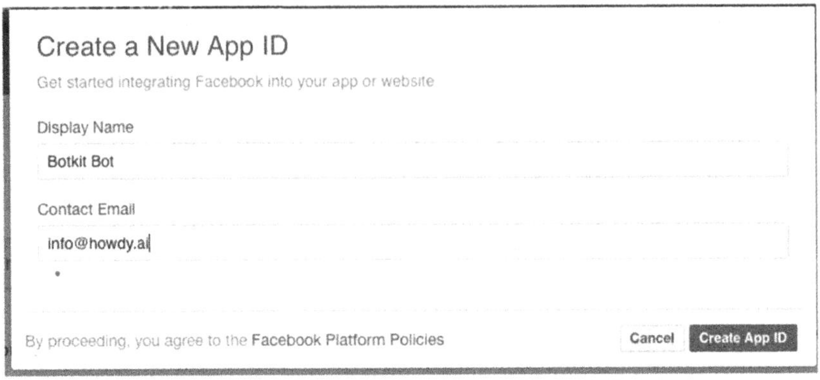

Figure 8-4. *Creation of new app ID on Facebook*

2. We created a Facebook page called AskIris and generated a page access token for the app. See Figure 8-5.

Figure 8-5. *Showing the page access token screen*

The callback URL is `https://YOURURL/facebook/receive`. See Figure 8-6. This URL must be publically available and SSL-secured. We provide a `localtunnel` callback URL; in our case, it's the NodeJs server URL tunneled to localhost. We will discuss how to create this endpoint and use of `localtunnel` in the next section.

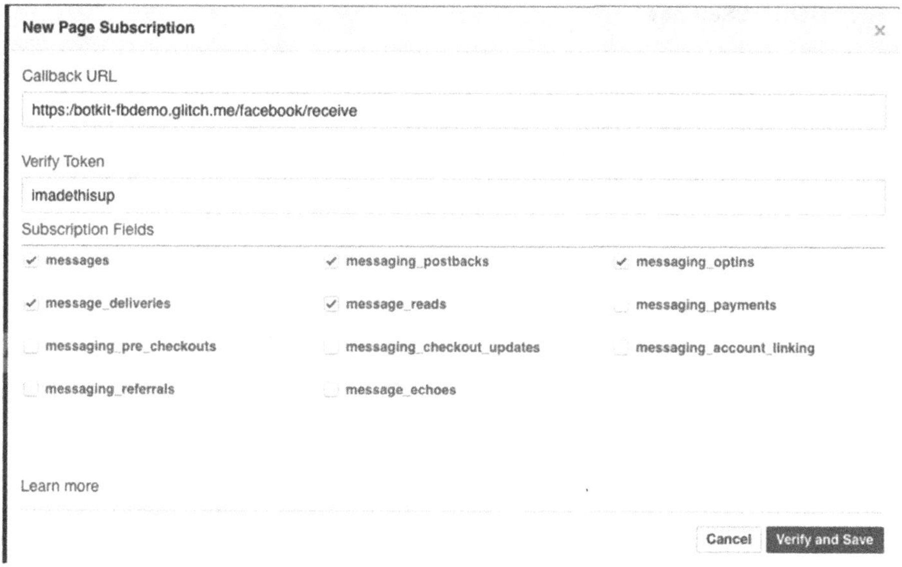

New Page Subscription   ×

Callback URL

https:/botkit-fbdemo.glitch.me/facebook/receive

Verify Token

imadethisup

Subscription Fields

✓ messages	✓ messaging_postbacks	✓ messaging_optins
✓ message_deliveries	✓ message_reads	messaging_payments
messaging_pre_checkouts	messaging_checkout_updates	messaging_account_linking
messaging_referrals	message_echoes	

Learn more

Cancel   Verify and Save

Figure 8-6. Showing page subscription details along with our callback URL

The AskIris page should be connected to the newly created app, and the settings should show the app details. See Figure 8-7.

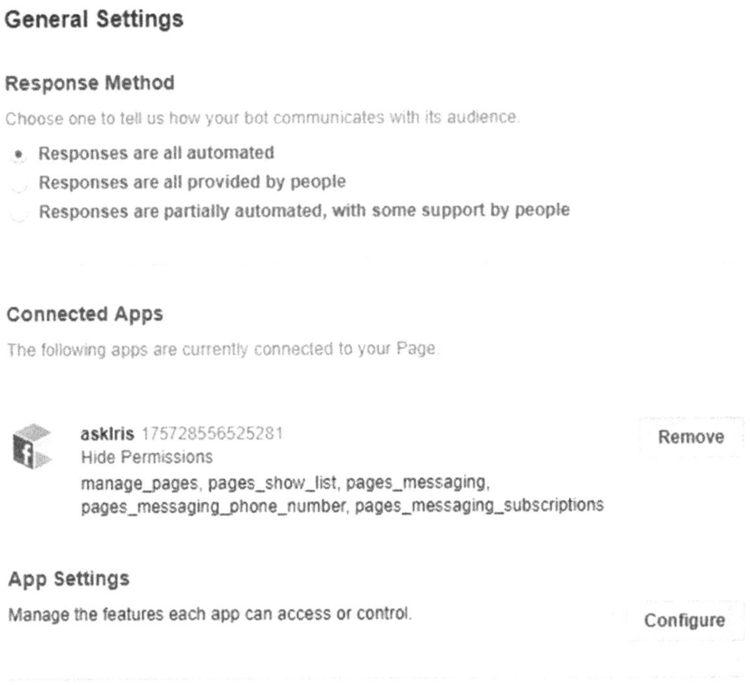

General Settings

Response Method

Choose one to tell us how your bot communicates with its audience.

- • Responses are all automated
- ○ Responses are all provided by people
- ○ Responses are partially automated, with some support by people

Connected Apps

The following apps are currently connected to your Page.

askIris 175728556525281
Hide Permissions
manage_pages, pages_show_list, pages_messaging,
pages_messaging_phone_number, pages_messaging_subscriptions

Remove

App Settings

Manage the features each app can access or control.

Configure

Figure 8-7. *The AskIris page settings*

As described above, setting up the webhook requires an HTTPS callback URL. This URL is the API endpoint of our integration module, which will receive messages from Messenger. We create a NodeJS application for this purpose because it is the requirement for setting up the webhook.

More details on setting up are described in Facebook developer page at

```
https://developers.facebook.com/docs/messenger-platform/
getting-started/webhook-setup.
```

Here are the steps to create a simple NodeJS application:

1. Create an HTTP Server (`server.js`):

   ```
   // Import modules required in server.js
   var express = require('express');
   var bodyParser = require('body-parser');
   ```

```
var https = require('https');
var http = require('http');
var fs = require('fs');
var localtunnel = require('localtunnel');

// Custom JavaScripts
var conf = require(__dirname + '/conf.js');
function server(ops) {

  // Create App
  /* Express is a popular Node web framework and
     provides a mechanism to write handlers.
  */
   var app = express();

  /* body-parser is Node.js middleware that parses
     incoming request bodies in a middleware before
     handlers.
  */
// parse JSON
   app.use(bodyParser.json());
//Returns middleware that only parses urlencoded bodies
   app.use(bodyParser.urlencoded({
//This object will contain key-value pairs, where
the value can be a string or array (when extended is
false), or any type (when extended is true).
       extended: true
   }));

// Path to static files
   app.use(express.static(__dirname + conf.static_dir));
/* Declare option and create a HTTPS server by reading
   SSL key and SSL cert path from config file.
*/
```

```
var options = {
    port : conf.securePort,
    key : fs.readFileSync(conf.sslKeyPath),
    cert : fs.readFileSync(conf.sslCertPath),
    requestCert : false,
    rejectUnauthorized : false
};
https.createServer( options, app)
.listen(conf.securePort,  conf.hostname, function()
{
    console.log('** Starting secure webserver on
    port ' + conf.securePort);
});

http.createServer(app)
.listen(conf.port, conf.hostname, function() {
    console.log('** Starting webserver on port ' +
    conf.port);
});

/*
localtunnel exposes localhost to the world for easy
testing and sharing. It will connect to the tunnel
server, set up the tunnel, and tell you what URL to use
for your testing. We used localtunnel to get an https
endpoint on http://localhost:9080/respond to test with
Facebook Messenger.
*/
  if(ops.lt) {
      var tunnel = localtunnel(conf.port, {subdomain:
      'askiris'}, function(err, tunnel) {
          if (err) {
```

```
                console.log(err);
                process.exit();
            }
            console.log("Your bot is available on the
            web at the following URL: " + tunnel.url +
            '/facebook/receive');
        });

        tunnel.on('close', function() {
            console.log("Your bot is no longer available
            on the web at the local tunnel.me URL.");
            process.exit();
        });
    }
    return app;
}
/* module.exports is an object that the current module
   returns when it is "required" in another program or
   module
*/
module.exports = server;
```

2. Add the Facebook webhook endpoints (Webhooks.js):

```
const fetch = require("node-fetch");

// This is the IRIS API URL endpoint
const url = "http://localhost:9080/respond";

function webhooks(controller){

/* This is the initial message a user sees before
   interacting with Iris for THE first time.
*/
```

```
controller.api.messenger_profile.greeting('Hi, my
name is IRIS. I am continuously training to become
your Digital Virtual Assistant\'s.');

// All messages will be sent to the API.
controller.hears(['.*'], 'message_received,facebook_
postback', function(bot, message) {
// Facebook request message contains text, senderID,
seq, and timestamp.
    var params = {
        message: message.text,
        sender: message.sender.id,
        seq: message.seq,
        timestamp: message. timestamp
    };

    var esc = encodeURIComponent;
    var query = Object.keys(params)
    .map(k => esc(k) + '=' + esc(params[k]))
    .join('&');

/* fetch makes a HTTP GET call and receives a response
    and passes the 'message' back to Facebook.
*/
    fetch(url +query)
    .then(response => {
        response.json().then(json => {
            bot.reply(message, json.message);
        });
    })
    .catch(error => {
        bot.reply(message, "");
    });
```

```
    });

}

module.exports = webhooks;
```

3. Add webhook verification. More details on specific
    of botkit-facebook integration are explained in detail
    at www.botkit.ai/docs/readme-facebook.html.

```
var Botkit = require('botkit');
var commandLineArgs = require('command-line-args');
var localtunnel = require('localtunnel');

// Reading static files
var server = require(__dirname + '/server.js');
var conf = require(__dirname + '/conf.js');
var webhook = require(__dirname + '/webhooks.js');

// Command line arguments to run in local mode vs
server; we need to use localtunnel to connect to
Facebook Messenger webhook in local mode as it requires
an HTTPS endpoint.
const ops = commandLineArgs([
    {name: 'lt', alias: 'l', args: 1, description:
    'Use localtunnel.me to make your bot available on
    the web.',
    type: Boolean, defaultValue: false},
    {name: 'ltsubdomain', alias: 's', args: 1,
    description: 'Custom subdomain for the
    localtunnel.me URL. This option can only be used
    together with --lt.',
    type: String, defaultValue: null},
]);
```

```javascript
// Create the Botkit controller, which controls all
instances of the bot.
var controller = Botkit.facebookbot({
    debug: true,
    log: true,
    access_token: conf.access_token,
    verify_token: conf.verify_token,
    app_secret: conf.app_secret,
    validate_requests: true
});
// Create server
var app = server(ops);

// Receive post data from FB; this will be the messages
you receive.
app.post('/facebook/receive', function(req, res) {
    if (req.query && req.query['hub.mode'] ==
    'subscribe') {
        if (req.query['hub.verify_token'] == controller.
        config.verify_token) {
            res.send(req.query['hub.challenge']);
        } else {
            res.send('OK');
        }
    }

    // Respond to Facebook that the webhook has been
    received.
    res.status(200);
    res.send('ok');

    var bot = controller.spawn({});
```

```
    // Now, pass the webhook to be processed.
    controller.handleWebhookPayload(req, res, bot);

});

// Perform the FB webhook verification handshake with
your verify token. The verification token is stored in
the conf.js file.
app.get('/facebook/receive', function(req, res) {
    if (req.query['hub.mode'] == 'subscribe') {
        if (req.query['hub.verify_token'] == controller.
        config.verify_token) {
            res.send(req.query['hub.challenge']);
        } else {
            res.send('OK');
        }
    }else{
        res.send('NOT-OK');
    }
});

// Ping URL
app.get('/ping', function(req, res) {
        res.send('{"status":"ok"}');
});

webhook(controller);
```

Once we are all done with the above and have tested the endpoints, we are all set to start interacting with IRIS on Facebook Messenger.

Demonstration of AskIris Chatbot in Facebook Messenger

Let's go through some example interactions. We can begin interacting with IRIS with a simple salutation to which IRIS replies with details about herself. See Figure 8-8.

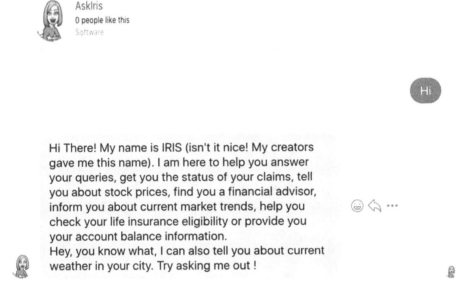

Figure 8-8. *Interaction with IRIS*

Account Balance

When the user asks IRIS for an account balance, it responds with a message asking for the confidential IPIN in order to proceed forward. As stated previously, demo implementation should not be adopted as a practice for setting a PIN. More complex and standard security authentication mechanism are available and should be followed. Upon a successful IPIN (which is a hard-coded value in the example use case), IRIS will retrieve the account balance for the account type the user asked for. See Figure 8-9.

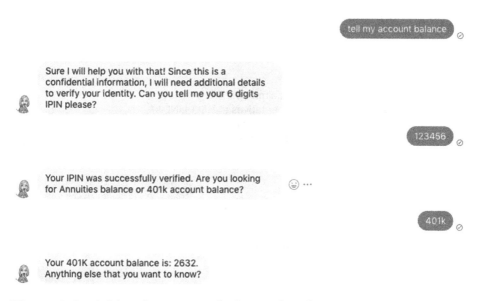

Figure 8-9. *Asking for account balance details*

Claim Status

The claim example demonstrates

- Identification of intent as well as slot required for intent fulfilment in one utterance. In example 1 below, intent (CLAIM_STATUS) and claim ID (gi123) are obtained at the same time.

- This demonstrates the potential of natural language processing to analyze natural user utterances, including spelling mistakes.

- Also, it shows the handling of variations in which the user asks for the same information. In example 1, we have natural language-based inference of intent and slot. In example 2, since the slot is not obtained, IRIS prompts for this information just like any other typical conversation. See Figure 8-10.

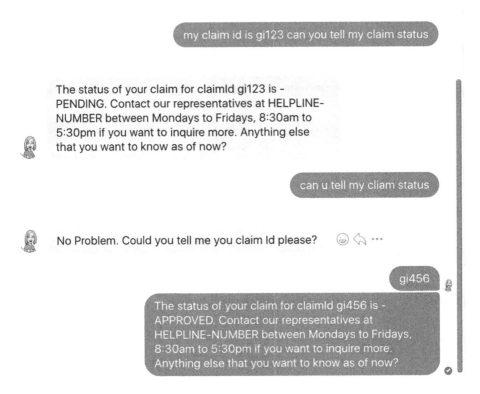

Figure 8-10. *Requesting claim status*

Weather Today

The weather example is a demonstration of how we can build a chatbot to integrate with third-party services in real time. User utterances such as the following return the live temperature information as available from the API (see Figure 8-11):

- What's the weather in Letterkenny today

- Dublin weather today

- Weather in Ranchi now

It seems to be broken clouds at the moment in letterkenny. The temperature is 9.0 degrees. Humidity is close to 87.
 I wish I were human to feel it. Anyways, what else do you want to know from me?

Figure 8-11. *Real time weather info from IRIS*

Frequently Asked Questions

First, you can see the correct response to an FAQ. Then, when a user utterance (401k) neither matches any explicit intent nor any document in the knowledge repository, IRIS performs a search. See Figure 8-12.

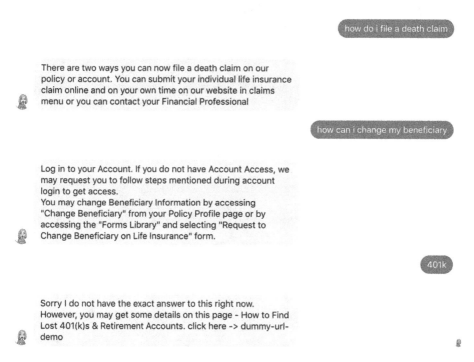

how do i file a death claim

There are two ways you can now file a death claim on our policy or account. You can submit your individual life insurance claim online and on your own time on our website in claims menu or you can contact your Financial Professional

how can i change my beneficiary

Log in to your Account. If you do not have Account Access, we may request you to follow steps mentioned during account login to get access.
You may change Beneficiary Information by accessing "Change Beneficiary" from your Policy Profile page or by accessing the "Forms Library" and selecting "Request to Change Beneficiary on Life Insurance" form.

401k

Sorry I do not have the exact answer to this right now. However, you may get some details on this page - How to Find Lost 401(k)s & Retirement Accounts. click here -> dummy-url-demo

Figure 8-12. *FAQ example*

Context Switch Example

In Figure 8-13, the user starts by asking for a life insurance quote. IRIS prompts for age, smoker info, height, and weight. The user provides all other information as expected except for weight. Instead of providing weight, user asks for a stock price. Instead of replying to the user that they have not entered the weight correctly or that their weight was not recognized, IRIS understands the switch in context and hence the switch in intent and seamlessly provides the requested detail. Later, when the user asks for life insurance again, notice that the already answered questions are not asked again. This is done because of short term memory in which these details were stored.

When the user provides the answer to the question that they didn't answer earlier, the intent is fulfilled and the response is obtained.

Figure 8-13. *Life insurance example illustrating context switch*

Summary

In this chapter, we showed how to extend the functionalities of IRIS to support integration with various third-party services. Connecting with an enterprise database to fetch user data was also explained. We discussed the integration module through which IRIS is exposed to the outside world. We implemented integration of IRIS with Facebook Messenger and followed the step-by-step process required for a successful integration.

Finally, we illustrated what interaction with Facebook Messenger and IRIS looks like. We showed a few use cases via examples and explained behind the scenes what was discussed in previous chapters.

In the next chapter, we will discuss deploying this in-house framework to the cloud. We will explore various ways in which we can deploy our services to AWS. We will also discuss how IRIS can be integrated with Alexa in less than 5 minutes by going through a step-by-step process. We will conclude by discussing how this framework can be improved and the scope for further enhancements such as the implementation of a feedback loop.

CHAPTER 9

Deployment and a Continuous Improvement Framework

In the previous chapters, we designed a basic chatbot framework from scratch and explored integration options with third-party services and other backend systems. We also explained how to expose the IRIS chatbot framework as a Spring Boot REST API.

In this chapter, we will discuss different ways in which IRIS can be deployed on a remote server. We will also discuss how to integrate IRIS with Alexa in less than 5 minutes. At the end of the chapter, we will discuss how IRIS can be extended to be part of a continuous improvement framework by implementing a self-learning module and bringing a human into the loop.

Deployment to the Cloud

The IRIS framework exposed via RESTful APIs can be deployed to a remote server in multiple ways. In this section, we will discuss three different ways.

© Abhishek Singh, Karthik Ramasubramanian, Shrey Shivam 2019
A. Singh et al., *Building an Enterprise Chatbot*,
https://doi.org/10.1007/978-1-4842-5034-1_9

As a Stand-Alone Spring Boot JAR on AWS EC2

This is the most basic installation and deployment of a Spring Boot JAR. We follow a few steps for the JAR to run on the EC2 machine on port 8080

We log into the AWS account and select EC2 from the list of services seen in Figure 9-1. EC2 stands for elastic compute cloud servers provided by AWS. More details on EC2 can be found at `https://aws.amazon.com/ec2/`.

Figure 9-1. *Different AWS services*

We launch an EC2 instance from the EC2 dashboard as shown in Figure 9-2.

Figure 9-2. *Launching an EC2 instance wizard on AWS*

Launching an EC2 instance requires seven steps:

1. We choose an AMI (Amazon Machine Image). We use Amazon Linux 2 AMI (HVM), SSD Volume Type 64 bit x86.

2.   We choose the instance type. We select **t2.micro**
     (also free tier eligible if you are using this service of
     AWS for the first time). The t2.micro instance has
     one vCPUs and 1 GB of memory, which is enough
     for the APIs to run.

3.   The next step requires configuring instance details.
     We can use a checklist to protect against accidental
     termination. This step is optional.

4.   We add storage details in the next step. By default,
     we get 8GB of SSD, and the volume is attached to
     the instance. However, we can add more volumes
     or increase the storage of the default volume if we
     want. This step is also optional, and 8GB of storage
     is enough for deployment for the demo.

5.   We add tags to instances and storage volume for better
     management of EC2 resources. This is also optional.

6.   This step, as shown in Figure 9-3, requires
     configuring a security group. A security group is a set
     of firewall rules that control the traffic for an instance.

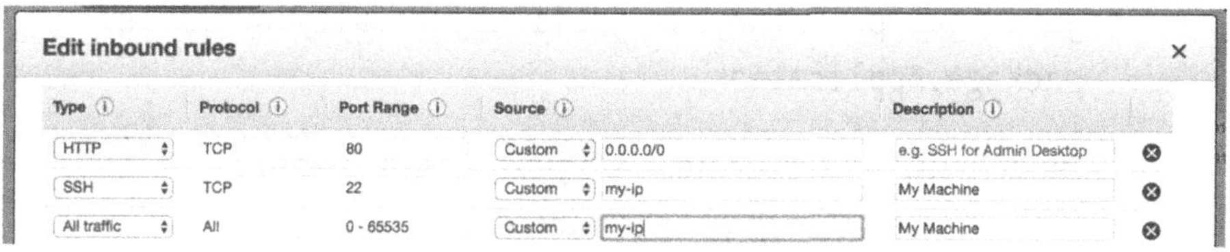

Figure 9-3. *EC2 security group inbound rules configuration*

We want to expose port 80 to be accessed from
everywhere and port 22, which is a secure shell port,
to be accessed only from our local machine.

7. We review the configuration and launch the instance. Each EC2 instance requires a key-pair PEM file that we need to log into the instance securely. We will be asked to generate a new file, or we can use an existing one.

Now once the instance is launched, it will have a public DNS name or IPv4 Public IP that we can use to log in.

1. The login command to log in from any Unix machine:

```
ssh -i chatbot-iris.pem ec2-user@ec2-instance-ip.
compute-1.amazonaws.com
```

2. Once we log in, we can then copy our Spring Boot JAR from local using the SCP command:

```
scp -i chatbot-iris.pem /path/to/iris.jar ec2-user@
ec2-instance-ip.compute-1.amazonaws.com:/path/to/your/
jarfile
```

3. Once the JAR is copied, we can run the JAR by issuing the command

```
java -jar path/to/your/jarfile.jar fully.qualified.
package.Application
```

4. By default, the server starts on port 8080. However, if we want to change the port details, we can set the server.port as a system property using command line options such as -DServer.port=8090 or add application.properties in src/main/resources/ with server.port=8090.

If we used maven to build our code, we could also use

```
mvn spring-boot:run
```

As a Docker Container on AWS EC2

Docker performs operating-system-level virtualization. Docker is used to run software packages called containers. Docker makes it easier from an operations perspective because it packages the code, libraries, and runtime components together as Docker images that can be deployed with a lot of ease. For more details on Docker, visit www.docker.com/.

We perform the following steps to run an application on Docker on EC2:

1. We update the installed packages and package cache on the instance:

    ```
    sudo yum update -y
    ```

2. We install the most recent Docker Community Edition package:

    ```
    sudo amazon-linux-extras install docker
    ```

3. We start the Docker service:

    ```
    sudo service docker start
    ```

4. We add the ec2-user to the Docker group in order to execute Docker commands without using

    ```
    sudo - sudo usermod -a -G docker ec2-user
    ```

5. We log out and log back in again to pick up the new Docker group permissions. To do so, we close the current SSH terminal window and reconnect to an instance in a new one. The new SSH session will have the appropriate Docker group permissions.

6. We verify that the ec2-user can run Docker
    commands without sudo.

7. We create a Dockerfile in the root directory of
    the code base. A Dockerfile is a manifest that
    describes the base image to use for the Docker
    image and whatever is installed and running on
    it. This dockerfile uses the openjdk:8-jdk-alpine
    image because we are building an image of a Java
    application. The VOLUME instruction creates a mount
    point with the specified name and marks it as holding
    externally mounted volumes from native host or other
    containers. The ARG instruction defines a variable that
    users can pass at build-time to the builder. The JAR is
    named as app.jar, and an ENTRYPOINT allows us to
    configure a container that will run as an executable. It
    contains the command to run the JAR:

```
FROM openjdk:8-jdk-alpine
VOLUME /tmp
ARG JAR_FILE
ADD ${JAR_FILE} app.jar
ENTRYPOINT ["java","-Djava.security.egd=file:/dev/./
urandom","-jar","/app.jar"]
```

8. We build a Docker image by issuing the following
    command:

```
docker build -t iris --build-arg JAR_FILE="JAR_NAME".
```

The following is the output from the build command
executed on a machine:

```
Sending build context to Docker daemon   21.9MB
Step 1/5 : FROM openjdk:8-jdk-alpine
```

```
8-jdk-alpine: Pulling from library/openjdk
bdf0201b3a05: Pull complete
9e12771959ad: Pull complete
c4efe34cda6e: Pull complete
Digest: sha256:2a52fedf1d4ab53323e16a032cadca89aac47024
a8228dea7f862dbccf169e1e
Status: Downloaded newer image for openjdk:8-jdk-alpine
 ---> 3675b9f543c5
Step 2/5 : VOLUME /tmp
 ---> Running in dc2934059ab8
Removing intermediate container dc2934059ab8
 ---> 0c3b61b6f027
Step 3/5 : ARG JAR_FILE
 ---> Running in 36701bf0a68e
Removing intermediate container 36701bf0a68e
 ---> da1c1f51c29d
Step 4/5 : ADD ${JAR_FILE} app.jar
 ---> 0aacdba5bafo
Step 5/5 : ENTRYPOINT ["java","-Djava.security.
egd=file:/dev/./urandom","-jar","/app.jar"]
 ---> Running in f40f7a276e18
Removing intermediate container f40f7a276e18
 ---> 493abfce6e8c
Successfully built 493abfce6e8c
Successfully tagged iris:latest
```

9. We run the newly created Docker image via the
    following command:

```
docker run -t -i -p 80:80 iris
```

As an ECS Service

In the previous two methods, you saw that you could deploy and run the Spring Boot JAR as a standalone service or by installing Docker and using it to run the API in the Docker container. This method discusses a service of AWS called ECS (Elastic Container Service). See Figure 9-4.

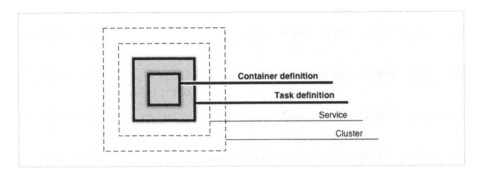

Figure 9-4. *Diagram of ECS objects and how they relate*

Amazon ECS makes it easy to deploy, manage, and scale Docker containers running applications, services, and batch processes. Amazon ECS places containers across your cluster based on your resource needs and is integrated with familiar features like elastic load balancing, EC2 security groups, EBS volumes, and IAM roles. More details on ECS can be found at https://aws.amazon.com/ecs/.

There are various steps required in running Docker images on ECS. Here's a walkthrough of the deployment process via the AWS management console:

1. When discussing how to deploy a JAR as a Docker container on AWS EC2, we created a Docker image. We need to add this previously created Docker image to ECR. Amazon Elastic Container Registry (ECR) is a fully-managed container registry that makes it easy for developers to store, manage, and deploy container images.

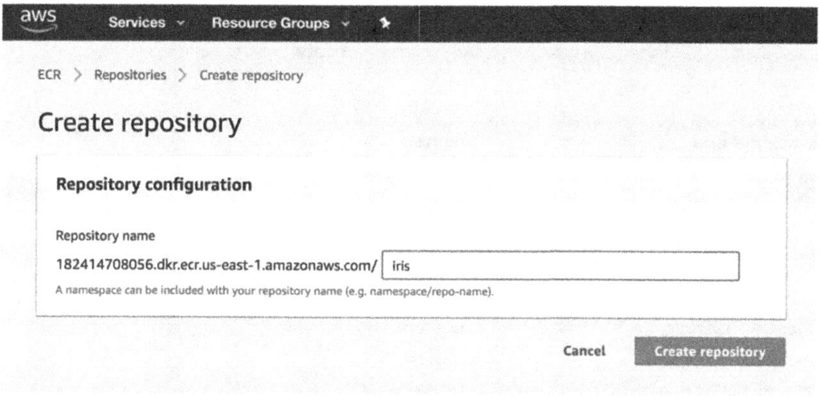

Figure 9-5. *Creating an ECR repository in AWS*

Once the repository is created as shown in Figure 9-5, the next step is to tag the Docker image so we can push the image to this repository:

```
docker tag iris:latest aws_account_id.dkr.ecr.us-east-1.amazonaws.com/iris:latest
```

Then we run the following command to push this image to ECR repository:

```
docker push aws_account_id.dkr.ecr.us-east-1.amazonaws.com/iris:latest
```

More details on pushing Docker images to ECR can be found at https://docs.aws.amazon.com/AmazonECR/latest/userguide/docker-push-ecr-image.html.

2. Then we need to define the container definition. In the ECS service in the AWS management console, under Get Started, we can choose a container definition to use. We need to provide the ECR repository URL and Docker image name and tag, as shown in Figure 9-6.

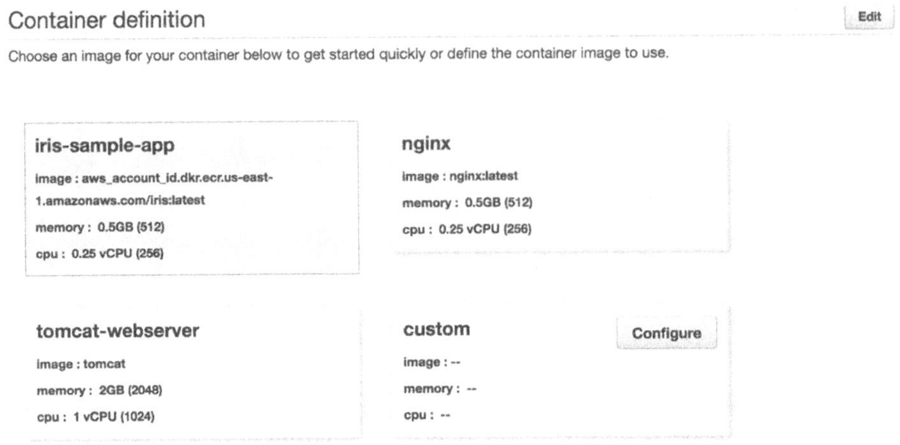

Container definition

Edit

Choose an image for your container below to get started quickly or define the container image to use.

iris-sample-app

image : aws_account_id.dkr.ecr.us-east-
1.amazonaws.com/iris:latest

memory : 0.5GB (512)

cpu : 0.25 vCPU (256)

nginx

image : nginx:latest

memory : 0.5GB (512)

cpu : 0.25 vCPU (256)

tomcat-webserver

image : tomcat

memory : 2GB (2048)

cpu : 1 vCPU (1024)

custom

image : --

memory : --

cpu : --

Configure

Figure 9-6. *Container definition configuration*

3. We define the task definition. A task definition is a
   blueprint for an application and describes one or
   more containers through attributes. Some attributes
   are configured at the task level, but the majority of
   attributes are configured per container. In Figure 9-7,
   we create a task definition for IRIS.

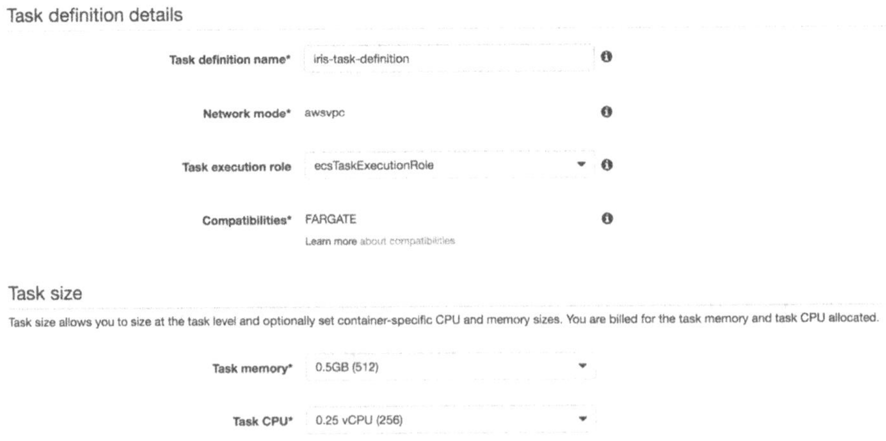

Task definition details

Task definition name*	iris-task-definition	ⓘ
Network mode*	awsvpc	ⓘ
Task execution role	ecsTaskExecutionRole	▾ ⓘ
Compatibilities*	FARGATE	ⓘ
	Learn more about compatibilities	

Task size

Task size allows you to size at the task level and optionally set container-specific CPU and memory sizes. You are billed for the task memory and task CPU allocated.

Task memory*	0.5GB (512)	▾
Task CPU*	0.25 vCPU (256)	▾

Figure 9-7. *Task definition*

4. Define a service. A service allows us to run and
    maintain a specified number (the desired count)
    of simultaneous instances of a task definition in an
    ECS cluster. See Figure 9-8.

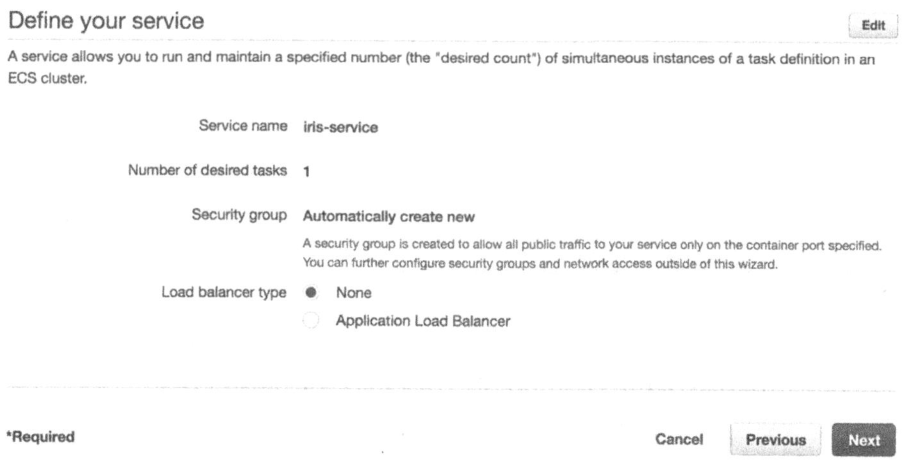

Figure 9-8. *Service definition details*

5. We configure a cluster. The infrastructure in a Fargate
    cluster is fully managed by AWS. Our containers run
    without us managing and configuring individual
    Amazon EC2 instances. See Figure 9-9.

Configure your cluster

The infrastructure in a Fargate cluster is fully managed by AWS. Your containers run without you managing and configuring individual Amazon EC2 instances.

To see key differences between Fargate and standard ECS clusters, see the Amazon ECS documentation.

Cluster name	iris-cluster
	Cluster names are unique per account per region. Up to 255 letters (uppercase and lowercase), numbers, and hyphens are allowed.
VPC ID	Automatically create new ⓘ
Subnets	Automatically create new ⓘ

*Required   Cancel   Previous   **Next**

Figure 9-9. *Cluster configuration details*

6. Once we review and click Create, we should see the progress of the creation of ECS. Once the cluster is set up and task definitions are complete, the Spring Boot service should be up and running.

Figure 9-10 is an example of a basic ECS cluster creation using the task definition outlined above.

Getting Started with Amazon Elastic Container Service (Amazon ECS) using Fargate

Launch Status

We are creating resources for your service. This may take up to 10 minutes. When we're complete, you can view your service.

Back   ○ View service   Enabled after service creation completes successfully

Additional features that you can add to your service after creation

Scale based on metrics
You can configure scaling rules based on CloudWatch metrics

Preparing service : 7 of 9 complete

ECS resource creation	pending ◌
Cluster iris-cluster	complete ⊘
Task definition iris-task-definition:1	complete ⊘
Service	pending ◌
Additional AWS service integrations	pending ◌
Log group /ecs/iris-task-definition	complete ⊘
CloudFormation stack	pending ◌
VPC vpc-05eb33683c927fb4c	complete ⊘
Subnet 1 subnet-0de45f5c768c39e5e	complete ⊘
Subnet 2 subnet-0629816622526533	complete ⊘
Security group sg-0f95bc581e87199db	complete ⊘

Figure 9-10. *The ECS Launch Status screen*

356

Smart IRIS Alexa Skill Creation in Less Than 5 Minutes

We will explore how IRIS can be integrated with Amazon Alexa in a few easy steps. In order to do so, the first step is to log into Alexa Developer Console at `https://developer.amazon.com/alexa` and create a skill. Creating a skill requires providing a skill name and the default language. We will choose a custom model to add to the skill. See Figure 9-11.

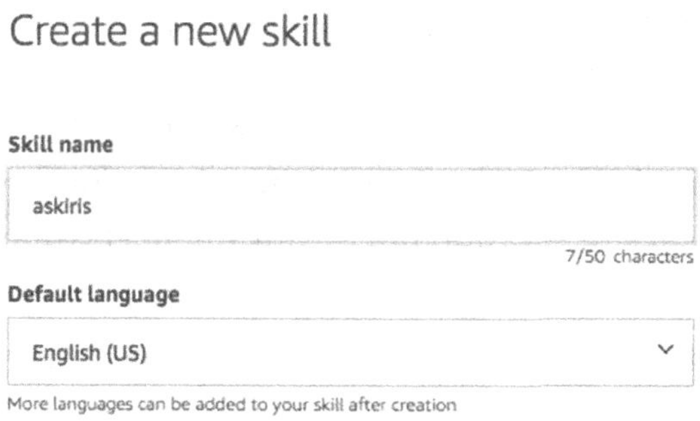

Create a new skill

Skill name

```
askiris
```
7/50 characters

Default language

```
English (US)                              ∨
```
More languages can be added to your skill after creation

Figure 9-11.   *Creating the Alexa skill name*

The custom model has a few checklists to be provided for the skill to work:

- Invocation name

- Intents, sample, and slots

- Building an interaction model

- Setting up a web service endpoint

In our example use case of IRIS, since we already have custom defined different possible intents, intent slots, and dialogs modeled as a state machine, we aim to redirect the user's utterance on Alexa to the IRIS backend API so that it can process the utterance and respond.

We use the invocation name as Iris. This will enable users to invoke this skill by asking Alexa to *"ask"* Iris. See Figure 9-12.

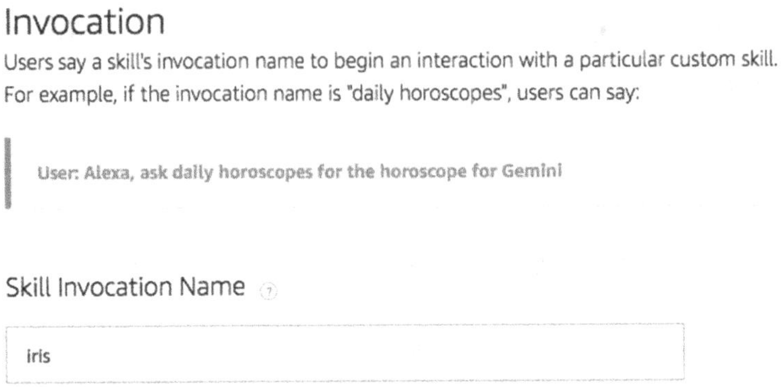

Invocation

Users say a skill's invocation name to begin an interaction with a particular custom skill. For example, if the invocation name is "daily horoscopes", users can say:

User: Alexa, ask daily horoscopes for the horoscope for Gemini

Skill Invocation Name ⓘ

iris

Figure 9-12.   Adding the skill invocation name

Next, we define a custom intent and a custom slot type so that all of the user's utterances are matched to this intent and slot type. The aim is to redirect the utterance to IRIS and not do any intent classification-related processing on the Alexa layer.

We first create a custom slot type called `IrisCustomSlotType`; see Figure 9-13.

Slot Types

| + Add Slot Type | Filter Slot Types | | |

NAME	SLOT VALUES	TYPE	ACTIONS
IrisCustomSlotType	1	Custom	Edit | Delete

Figure 9-13.   A custom slot type

Now, we define a custom intent named as `IrisAllIntent`. This intent has a slot called `utteranceSlot`. The most important thing is `{utteranceSlot}`. We intend to fit all user utterances into this slot, as shown in Figure 9-14. It is a regex, which means the entire utterance value is under the `utteranceSlot` slot. This will be used later while reading the user utterance when Alexa requests the IRIS HTTPS endpoint.

Intents / irisAllIntent

Sample Utterances (1) ⓘ   🗑 Bulk Edit   ⬇ Export

| What might a user say to invoke this Intent? | + |

| {utteranceSlot} | 🗑 |

< 1 – 1 of 1 >

Figure 9-14. *The utteranceSlot in IrisAllIntent*

The `utteranceSlot` is defined to be of slot type `IrisCustomSlotType`, as shown in Figure 9-15.

Figure 9-15. *Custom slot and its type created for IRIS integration*

At this stage, we should have the intents, slots, and slot types created in the interaction model of our Alexa skill, as shown in Figure 9-16. In the figure, you can see that there are other built-in intents as well what is present for a standard interaction with Alexa devices such as stop or cancel.

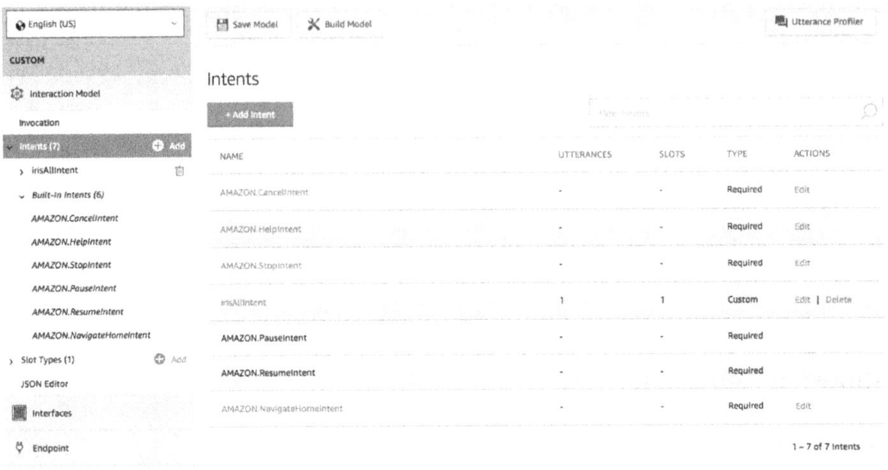

Figure 9-16. Interaction model screen

Once we have defined all of the required attributes for our custom interaction model, we can build the model. Building the model requires us to click the Build Model button, which saves the skill manifest and builds the model. See Figure 9-17.

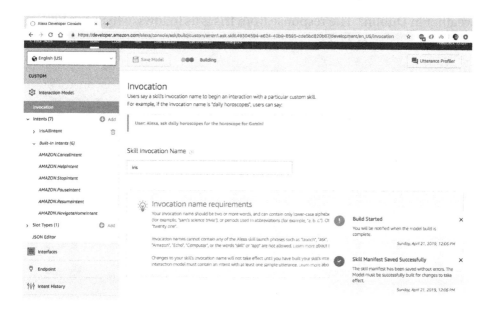

Figure 9-17. *The progress of the build process of the IRIS custom interaction model*

The last thing to complete the setup is to provide an HTTPS endpoint of IRIS to host all the intent classifications and response generation logic, and receive a POST request from Alexa. See Figure 9-18.

Service Endpoint Type

Select how you will host your skill's service endpoint.

○ AWS Lambda ARN ⓘ
(Recommended)

● HTTPS ⓘ

Default Region ⓘ
(Required)

https://bot.askiris.today/alexa

My development endpoint is a sub-domain of a domain that has a wildc... ˅

Figure 9-18. *The skill's service endpoint configuration*

Let's now test the setup by using the simulator available on the developer console, as shown in Figure 9-19. We ask Alexa to ask Iris for "weather in Dublin." The response from Iris is as follows, which can be heard in the voice of Alexa.

Figure 9-19. *The simulator available on the Alexa developer console*

The request details sent from Alexa to the IRIS API are

```
{
    "version": "1.0",
    "session": {
        "new": true,
        "sessionId": "amzn1.echo-api.session.XXXXXXX-2b66-
                XXXX-XXX-XXXXXXXXXX",
        "application": {
            "applicationId": "amzn1.ask.skill.XXXXXX-
                        e624-XXXXXX-XXX-XXXXXXX"
        },
        "user": {
            "userId": "amzn1.ask.account.XXXXXXXX"
        }
    },
```

```
"context": {
    "AudioPlayer": {
        "playerActivity": "IDLE"
    },
    "System": {
        "application": {
            "applicationId": "amzn1.ask.skill.
                             XXXXXXX-XXX-XXXXX-
                             XXXX-XXXXXXXXX"
        },
        "user": {
            "userId": "amzn1.ask.account.XXXXXX"
        },
        "device": {
            "deviceId": "amzn1.ask.device.XXXXX",
            "supportedInterfaces": {
                "AudioPlayer": {}
            }
        },
        "apiEndpoint": "https://api.eu.amazonalexa.
                       com",
        "apiAccessToken": "XXXXXXXX"
    },
    "Viewport": {
        "experiences": [
            {
                "arcMinuteWidth": 246,
                "arcMinuteHeight": 144,
                "canRotate": false,
                "canResize": false
            }
        ],
```

```
                "shape": "RECTANGLE",
                "pixelWidth": 1024,
                "pixelHeight": 600,
                "dpi": 160,
                "currentPixelWidth": 1024,
                "currentPixelHeight": 600,
                "touch": [
                        "SINGLE"
                ]
            }
    },
    "request": {
            "type": "IntentRequest",
            "requestId": "amzn1.echo-api.request.XXXXXXXXXXX",
            "timestamp": "2019-04-21T11:22:56Z",
            "locale": "en-US",
            "intent": {
                    "name": "irisAllIntent",
                    "confirmationStatus": "NONE",
                    "slots": {
                            "utteranceSlot": {
                                    "name": "utteranceSlot",
                                    "value": "weather in Dublin",
                                    "resolutions": {
                                        "resolutionsPerAuthority":
[
                                            {
```

```
                                "authority":
                                "amzn1.er-
                                authority.echo-
                                sdk.amzn1.ask.
                                skill.XXXXXXXX-
                                XXXX-XXXX-
                                XXXX-XXXXXXX.
                                IrisCustom
                                SlotType",
                                "status": {
                                        "code":
                                        "ER_
                                        SUCCESS_
                                        NO_MATCH"
                                }
                        }
                    ]
                },
                "confirmationStatus": "NONE",
                    "source": "USER"
                }
            }
        }
    }
}
```

Response from IRIS:

```
{
    "body": {
        "version": "string",
        "response": {
            "outputSpeech": {
```

```
                        "type": "PlainText",
                        "text": "It seems to be clear sky at
                                the moment in dublin. The
                                temperature is 19.13 degrees.
                                Humidity is close to 48.\n
                                I wish I were human to feel
                                it. Anyways, what else do you
                                want to know from me? "
                },
                "type": "_DEFAULT_RESPONSE"
        },
        "sessionAttributes": {}
    }
}
```

In Chapter 6, we created a NodeJS application that provides an external API endpoint for Facebook Messenger. For integrating with Alexa, we need to add the Alexa endpoint as described below:

```
// new API endpoint 'alexa' that expects a POST request
app.post('/alexa', function(req, res) {

// receive the user's utterance by reading the utteranceSlot
value from JSON

    var text = req.body.request.intent.slots.utteranceSlot.value;
    var session = req.body.session.user.userId;
    var timestamp = req.body.request.timestamp;

/* GET request parameters to the IRIS backend service.
'Message' param contains user utterance
*/
    var params = {
        message: text,
```

```
      sender: session,
      seq: 100,
      timestamp: 1524326401
   };

   var esc = encodeURIComponent;
   var query = Object.keys(params)
                  .map(k => esc(k) + '=' + esc(params[k]))
                  .join('&');
//url is the IRIS API url
   fetch(url +query).then(response => {
      response.json().then(json => {

         var alexaResp = {
             "version": "string",
             "sessionAttributes": {},
             "response": {
                 "outputSpeech": {
                     "type": "PlainText",
                     "text": json.message,
                     "ssml": "<speak>"+json.message+"</speak>"
                 }
             }
         }

         res.json(alexaResp);
         });
      })
      .catch(error => {

      var alexaResp = {
          "version": "string",
          "sessionAttributes": {},
```

```
    "response": {
        "outputSpeech": {
            "type": "PlainText",
            "text": "Sorry, My team is having bad day
            to get this information to you. Please try
            again in some time.",
            "ssml": "<speak>Sorry, My team is having bad
            day to get this information to you. Please
            try again in some time.</speak>"
        }
    }
}

res.json(alexaResp);
});

});
```

The details on hosting a custom skill as a web service are available at
`https://developer.amazon.com/docs/custom-skills/host-a-custom-skill-as-a-web-service.html`.

Continuous Improvement Framework

In practical cases, it is very possible that a user's utterances are not
classified or understood by our intent engine due to several reasons such
as the utterance being an external intent not part of the intent engine or
the intent engine not confident due to the low intent match score. In a
production environment, it is observed that there are a decent number
of user utterances that are either misunderstood or not understood by
the engine at all. We propose a framework that can help IRIS to become
smarter and more intelligent towards mimicking a natural human
conversation.

In the self-learning module, we propose three improvement components, shown in Figure 9-20:

- Intent confirmation (double-check)

- Next intent prediction

- A human in the loop

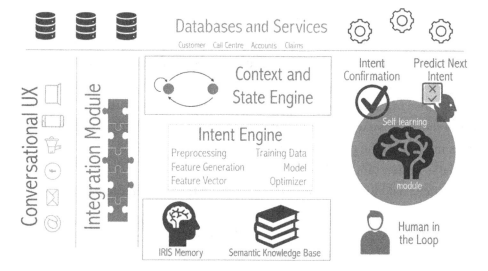

Figure 9-20. *IRIS functional components for continuous improvement*

Intent Confirmation (Double-Check)

Let's take an example of a user's utterance of "life insurance," which may match with one of the possible intent scenarios; see Figure 9-21.

POSSIBILITIES	Match Score
LIFE_INSURANCE_QUOTE_INTENT	0.85
BUY_LIFE_INSURANCE_INTENT	0.65

Figure 9-21. *Intent matches and corresponding scores*

When we match the user utterance against the list of possible intents shown in Figure 9-21, we get a list of intents and respective match scores. The intent engine module of IRIS returns with an intent match only when the match score is above 0.75. We also call this as the minimum threshold score below which an intent match is not considered in response. In the example of "life insurance," `LIFE_INSURANCE_QUOTE_INTENT` is returned in response from the intent engine.

An optimization to this implementation could be to introduce a minimum match score that is below the threshold score but relevant enough for further processing. We previously stated that the minimum threshold score is the score below which an intent match is not returned in response from the intent classification engine. A minimum match score is the score above which an intent is considered for further processing if it does not match the minimum threshold score.

Let's understand with another example of a user utterance: "life insurance cost," for which the match score is shown in Figure 9-22.

POSSIBILITIES	Match Score
LIFE_INSURANCE_QUOTE_INTENT	0.70
BUY_LIFE_INSURANCE_INTENT	0.45

Figure 9-22. *Another example showing different intent matches and corresponding scores*

In this example, the scores are below the minimum threshold score and in the current implementation the user utterance will default to search since no explicit intent was returned by the intent classification engine. If we consider a minimum match score of 0.5, intent `LIFE_INSURANCE_QUOTE_INTENT` could be considered for further confirmation.

These scores of 0.75 (minimum threshold score) and 0.5 (minimum match score) should be derived from training and test datasets, and could also change later on based on actual user utterance data and performance of intent classification engine.

Hence, we could make changes in IRIS to prompt for confirmation if the utterance was classified between 0.5 to 0.75.

With the above implementation, a typical conversation can then flow in the following ways:

```
User: Hello
IRIS: Hello
User: life insurance cost
IRIS: I could not understand your query correctly.  Do you want
      a life insurance quote?
User: Yes
IRIS: Sure, I can help you with that. May I know your age?
```

Predict Next Intent

This functionality enables IRIS to use logs to understand and predict the possible next best intent of the user. Figure 9-23 shows user logs.

User	Sequence 1	Sequence 2	Sequence 3	Sequence 4
User 1	INTENT-A	INTENT-B	INTENT-D	INTENT-E
User 2	INTENT-F	INTENT-D	INTENT-A	INTENT-B
User 3	INTENT-C	INTENT-E	INTENT-B	INTENT-D
User 4	INTENT-E	INTENT-C	INTENT-A	INTENT-D

Figure 9-23. *User logs with intent and sequence*

In a simple summarization of INTENT-A, shown in Figure 9-24, we find that INTENT-B is the most frequent next intent overall.

Intent	Next Intent sequence	Count
INTENT-A	INTENT-A	0
INTENT-A	INTENT-B	2
INTENT-A	INTENT-C	0
INTENT-A	INTENT-D	1
INTENT-A	INTENT-E	0
INTENT-A	INTENT-F	0

Figure 9-24. *Log summarization to predict next intent sequence*

We can also use techniques of path prediction, association rules, and frequent itemsets to obtain the most predictable next user intent.

A typical conversation with IRIS without the predicted intent may happen as follows:

```
User: Hi
IRIS: Hello
User: I want a life insurance quote
IRIS: Sure, I will help you with that. May I know your age?
User: My age is 24
IRIS: Have you smoked in the last 12 months?
User: no
IRIS: What's your height?
User: my height is 181 cms
IRIS: What's your weight?
User: 163 pounds
IRIS: Great news! You are eligible for an accelerated UW decision.
User: Thanks. I would like to buy life insurance policy.
IRIS: Sure. Can you provide me your email ID so that I start with
      the application process?
User: user@gmail.com
....
```

With predicted next intent capability, this conversation would happen like:

```
User: Hi
IRIS: Hello
User: I want a life insurance quote
IRIS: Sure, I will help you with that. May I know your age?
User: My age is 24
IRIS: Have you smoked in the last 12 months?
User: no
IRIS: What's your height?
User: my height is 181 cms
IRIS: What's your weight?
User: 163 pounds
IRIS: Great news! You are eligible for an accelerated UW
      decision. May I know your email ID so that I can start
      with the application process?
User: user@gmail.com
....
```

A Human in the Loop

The third improvement to the framework we introduced in this chapter was a human in the loop. Figure 9-20 shows various functional components for continuous improvement. Regardless of the techniques we use in making IRIS understand intent in a better way, there will always be some conversation that IRIS will not be able to understand. This is for the simple reason that IRIS does not have all the information of the universe and will always be designed to fulfill only a known set of functionalities.

We know that IRIS is designed to perform certain operations like calculating insurance eligibility, providing account balance, claim status, etc. Let's assume that a certain percentage of users are asking IRIS for a change of address of their insurance policy. This is not supported by IRIS today, and it is challenging for machines to interpret this kind of new information.

Let's assume some of the user utterances are as shown in Figure 9-25.

Utterance	Count
I want to change my address	7
Change address of insurance policy	5
What is brexit	2
When will my address change reflect in my policy?	4
Life insurance verify eligibility	3

Figure 9-25. *Utterances and counts*

With a human in the loop, these utterances can be analyzed further. In Figure 9-26, the first two utterances can be classified into one intent. Since this intent is 66% of the total logs, IRIS can be enhanced to support ADDRESS_CHANGE depending on the product decision. In some cases, an utterance can be mapped to an existing intent such as the last utterance. This will help the intent classification engine to further classify intents better due to the availability of more dataset.

Utterance	Intent	Count
I want to change my address	ADDRESS_CHANGE	7
Change address of insurance policy	ADDRESS_CHANGE	5
What is brexit		2
When will my address change reflect in my policy?	FAQ	4
Life insurance verify eligibility	LIFE_INSURANCE_QUOTE_INTENT	3

Figure 9-26. *Utterances, intents, and counts*

Utterances such as questions around non-related things such as users asking about cricket match scores or details on Brexit or train timings will also happen. They are logs that do not need further processing and will be ignored by subject matter experts enhancing the IRIS feedback loop.

Summary

In this concluding chapter, we discussed the various ways to deploy a chatbot into the cloud, we demonstrated a 5-minute introduction to integrating IRIS with Alexa, and we discussed the continuous improvement of IRIS through feedback loops via log files and humans in the loop.

In this book, we have kept a fine balance with three pillars: business context, theoretical foundations on machines' handling of natural languages, and real-world development of a chatbot from scratch. We believe these three pillars will help build a truly enterprise-grade chatbot, with a well-defined ROI. Additionally, we also focused on ethical concerns in using personal data and how countries in European Union have agreed upon the GDPR regulations to safeguard people's privacy.

Index

A

AccTypeSlot, 239–240
addTransition method, 265
Advent of technology, 3
AphaNumericSlot, 241
Architecture, private chatbots
    key features, 67, 68
    maintenance, 68, 69
    technology stack, 68
    workflow, 64–67
AskIris Chatbot, Facebook
    Messenger
    account balance, 338, 339
    claim status, 339, 340
    IRIS, 338
    life insurance, 342, 343
    weather, 340, 341
Authentication, 39
Authorization, 39

B

Bag-of-words (BOW), 139, 173, 174
Banks and insurers
    chatbot build process, 5
    commercial transactions, 2
    lenders and borrowers, 4
    mortality, 3

risk of loss, 2
theoretical framework, 2, 4
types of banks, 5
Bounce rate (BR), 61

C

CategorizedPlaintextCorpusReader
    method, 180
Chatbots, 73
    agents/advisors, insurance, 13
    AI-based approach
        advantages, 46
        disadvantages, 47
        NLP engine, 45, 46
    vs. apps, 56
    architecture, 64, 73, 74 (*see also*
        Architecture, private
        chatbots)
    automated underwriting, 12
    benefits, business, 58
        cost savings, 58, 59
        customer experience, 59, 60
    business, 9, 10, 55
    channel, 51
    conversations, 20–21, 42
    vs. direct contact, 57
    entity, 50

© Abhishek Singh, Karthik Ramasubramanian, Shrey Shivam 2019
A. Singh et al., *Building an Enterprise Chatbot*,
https://doi.org/10.1007/978-1-4842-5034-1

M

N

Printed in the United States
By Bookmasters